In Data Time and Tide

Cosimo Accoto

Foreword by Alex 'Sandy' Pentland
Afterword by Derrick De Kerckhove

B
U
P

A Surprising Philosophical Guide
to our Programmable Future

Original Title: *Il mondo dato*
Copyright © 2017 EGEA S.p.A.

Translator: Derrick De Kerckhove

Typesetting: Laura Panigara, Cesano Boscone (MI)

Copyright © 2018 Bocconi University Press
EGEA S.p.A.

EGEA S.p.A.
Via Salasco, 5 - 20136 Milano
Tel. 02/5836.5751 – Fax 02/5836.5753
egea.edizioni@unibocconi.it – www.egeaeditore.it

First edition: July 2018

ISBN Domestic Edition   978-88-99902-30-8
ISBN International Edition   978-88-85486-62-1
ISBN Moby pocket Edition   978-88-85486-64-5
ISBN Epub Edition   978-88-85486-63-8

Contents

Foreword

by *Alex 'Sandy' Pentland*

Magic

This book, written by Cosimo Accoto, describes how our culture and the concepts we use to know it will change as our world fills up with code, data, objects and platforms with a computational intelligence. One of the most profound changes is that we will move from speculatively asking "what are the facts?" to "what is going to happen?". Instead of retrieving dead documents talking about the past, we move to a future orientation where our code and software queries cause millions of scenarios to be simulated in order to foretell the future, as is already done today for weather, automobile traffic, and financial planning. To support these oracular powers, the world is also filling up with sensors and algorithms providing the raw data for accurate projection of the future.

As the world becomes alive with sensors and all objects have a computational intelligence we may become like magicians: we will be able to say "make it so" and the change, however complex, will happen...and often even in anticipation of our desires, so that seas of complexity and problems will part before us without our even being aware and experience so much has been done on our behalf mainly on a subperceptual, automatic, preemptive way.

* Alex Pentland is Professor at the Massachusetts Institute of Technology, member of U.S. National Academy, Advisor to U.N. Secretary General's office, leader in the World Economic Forum, and co-creator of the MIT Media Lab. He is among the most-cited academics in the world, and author of Social Physics (Penguin Press) and Honest Signals (MIT Press).

This vision of the future may seem to depend on an overly optimistic best-case future, but it applies equally to many dystopian futures. Moreover, and importantly, our culture is already being altered by the spread of invisible software, sensor data and computational intelligence. We no longer worry about getting lost or finding the best route, due to data concentrated from billions of cell phones. We no longer need to visit a physical travel agent to obtain precious paper tickets, our travel reservation is just there when we need it. And so on, and so on. But it is changing our culture so slowly that it is only people like Cosimo who can explore and philosophically see the big picture clearly.

Thinking about ourselves: from generalizations to predictions

As we move into a world made from data, most of the ways we think about the world, society and human behavior change in a rather dramatic way. For instance, philosophers like Adam Smith and Karl Marx never had more than half the answers. Why? Because they talked about markets and classes, but those are aggregates. They're averages.

While it may be useful to reason about the averages, social phenomena are really made up of millions of small transactions between individuals. There are patterns in those individual transactions that are not just averages, they're the things that are responsible for the financial crash of 2008 and the Arab spring. You need to get down into these myriad details, these micro-patterns, because they don't just average out to the classical way of understanding society. We're entering a new era of social physics, where it's the details of all the particles – the you and me – that actually determine the outcome.

Reasoning about markets and classes may get you half of the way there, but it's this new capability of looking at the details, which is only possible through ubiquitous data, that will give us the other 50 percent of the story. We can potentially design companies, organizations, and societies that are more fair, stable and efficient as we get to really understand human physics at this fine-grain scale. This new computational social science offers incredible possibilities – and powers that can be used both for good and bad.

This is the first time in human history that we have the ability to see enough about ourselves that we can hope to actually build social systems that work qualitatively better than the systems we've always had. Corporeal sensors, societal sensors and environmental sensors will offer us the opportunity to move from reality mining to reality making. The living labs we have set up, systems which allow entire communities of people to experiment with new rules and new capabilities, are a key invention that allows us to test and deploy ideas and determine if they actually lead to better futures or whether they have unintended consequences. That's a remarkable change, where changes in our society are no longer determined by speculative debate but instead are subjected to experimental evaluation. It's like the phase transition that happened when writing was developed or when education became ubiquitous, or perhaps when people began being tied together via the Internet. As happened for social physics, data – Cosimo writes – help us to reimagine other philosophical concepts such as: time, space, agency, subjectivity, law and experience.

The fact that we can now begin to actually look at the dynamics of social interactions and how they play out, and are not just limited to reasoning about averages like market indices is for me simply astonishing. To be able to see the details of variations in social outcomes and the beginnings of political revolutions, to predict them, and even control them, is definitely a case of Promethean fire. A world built on data can be good or bad, but either way it brings us to interesting times. We're beginning to reinvent what it means to have a human society.

A world of sensors and an emerging collective intelligence

The most powerful part of the human intelligence is our social brain, our ability to remember people, interactions and relationships. Indeed, the large size of the human brain seems mainly to be due to the need to keep track of all this social information. However, our society has not built many tools to support our social brain. Facebook, LinkedIn and other platforms are mostly either gossip machines or catalogs of resumes, run more for the benefit of the owners than the users.

But imagine that we could supercharge our social brain – giving people data-driven tools and anticipatory information that allow them to really know what is going on in companies, cities and governments.

To accomplish this supercharging, we need tools that aid our social brain through social sensing and relationship tuning, just as today's computer tools extend our memories and ability to calculate. By teaching computers more about how humans interact best, they can play the role of social secretaries and social network connectors. Algorithms, artificial intelligence and new platforms (if properly designed and openly shared) could enhance our humanity. A social data sense will hopefully improve our society, and we are building tools that let us test claims about the performance of new social systems.

The ability to find new ideas and create new connections is the bedrock for building human organizations that are creative and fast-moving. To understand how this works, think of an organization as a sort of brain, with the employees or members as the individual neurons. Static organizations – symbolized by the ubiquitous "org chart" – have fixed connections and, as a result, a limited ability to learn. Typically, these types of organizations become siloed, with little communication between departments, and cut off from new ideas. In that state, they risk falling to the competition. By supercharging the organizations' "social brain", the connections – between employees, departments and teams – can continuously reorganize themselves in response to shifting circumstances.

Importantly, this idea of adaptable connections is exactly the insight powering today's cutting-edge artificial intelligence, including both statistical machine learning and deep learning "neural net" approaches. In these models – as Cosimo clarifies – the connections between simple logic machines are reconfigured as the system learns. In contrast to logic machines, people can remake not just their connectivity but also their function, offering a fluid architecture that is qualitatively more powerful. Armed with the right feedback, human "smart neurons" can fill communication gaps to accelerate learning, anticipate "unknown unknowns" and invent new structures to address emerging market forces.

Dissolving boundaries: distributing power

As our world fills up with data, sensors, algorithms and objects with computational intelligence you can begin to create a new social sensorium, an entirely different world in which personal needs and desires are central, and rigid, machine-like uniformity fades away. This sort of oracular vision is already allowing us to engineer transportation, energy, and health systems that are dramatically personalized and consequently much better. We are at a phase transition. We are moving from the reasoning of the enlightenment about classes and about markets to fine-grain understanding of individual interactions and systems built to support the intentions and needs of individuals based on fine-grain data.

Importantly, the most secure and efficient data architectures are those that have no central points. In such distributed systems there's no single place for a dictator to grab control. So, security in a data rich society also means a higher level of transparency and choice for individuals, which mitigates against central control. New technological protocols such as blockchains may help to build new open and secured peer networks. The power of the state and big organizations tends to dissolve in a distributed data and computation rich world – Cosimo explains how new sovereignty regimes are emerging – because the organizations that will survive will be distributed among many stakeholders and without the hard information boundaries that you see today.

Magic for whom?

One of the great questions is: who is this new data rich world going to be for and what is it going to look like? A key insight is that your data is worth more if you share it because it enables systems like public health to work better for *you*. Data about the way you behave and where you go can be used to can stop the spread of infectious disease. If you have children, you don't want to see them die of an H1N1 pandemic. How are you going to stop that? Well, it turns out that if you can actually watch people's behavior in real time, something that is quite possible today, you can tell when each individual person is getting sick. This means you can actually see the spread of influenza from person to person on an individual level. And if you can see it, you can stop it. You can begin to build a world where infectious pandemics cease to be as much of a threat.

Similarly, if you're worried about global warming, we now know how patterns of mobility relate to productivity. This means you can design cities that are far more efficient, far more human, and burn an awful lot less energy. But you need to be able to see the people moving around in order to be able to get these results. That's another instance where sharing your data is invaluable to you personally. It's everybody contributing his or her data that's going to make a greener world, and that is worth far more than the simple cash value of the data.

But, of course, these examples assume that we have already put in place the correct governance, transparency, privacy, and accountability for such data and the consequent decisions. Moreover, today the data is often siloed off and unavailable for public use, and sharing personal data is dangerous because of data theft and bad actors. Vulnerability is, of course, an emergent property of our complex sociotechnical systems, as Cosimo makes clear. It was for these reasons that I proposed the New Deal on Data to the World Economic Forum in 2008. The New Deal is simple: people have rights to control data about them. Since the initial discussions in Davos the idea has run through various forums and turned into the Consumer Data Bill of Rights in the United States, and the Data Rights rules in the European Union (EU). The core idea is that people are willing to share their data if they can expect that it is safe, and they can derive personal benefit from sharing. We have also to digitally redesign the concept of our identities to better fit a digital world. Consequently, by giving power to individuals to control data about themselves, we can have the sort of democratized data-sharing environment that will allow us to create a healthier, greener and more peaceful world. The battle for personal privacy still rages, of course, but I believe that the tide has now turned in favor of the individual.

Hello, New World

Expressive code is creative, generative, and world-building
E. Swanstrom, *Animal, Vegetable, Digital*

As I am about to end this book, menacing clouds are hanging over our planet. The map I'm observing shows the areas where these turbulences have been most concentrated recently. However, these are not alarming atmospheric phenomena, although they have all the appearances of such. In a climate of mutual accusations and retaliations between the United States and Russia, a cyberattack was launched last year against the American internet infrastructure by suspected Russian hackers. The map view shows, like cyclones dropping on cities, the geographic points of the network that have been the subject of this massive information assault, the largest in the era of the internet. Because of these three cyberterror waves, a series of system downs prevented access to services and applications to millions of citizens, users and consumers. Particularly heavy was the attack on Dyn, one of the US hosting giants on which many internet companies rely, such as CNN, New York Times, Netflix, Twitter, Spotify, eBay and Visa, rendering unreachable those and several other platforms for about two hours.

This was an attack on network logistics (not directly on platforms), perpetrated – as far as it is known from the first reconstructions – through a *botnet*, a network of machines and objects of the internet of things. Millions of connected devices, such as webcams, thermostats and infected printers, were transformed into bots and used for a denial-of-service (DoS) attack aimed at preventing the use of resources and services on the network. In a direct response, American hackers – according to NBC –

penetrated the Kremlin's command-line information system, thus demonstrating their vulnerability.

And these are not the only episodes. In recent months, other hackers (assumed in this case to be Chinese) attacked an American aircraft carrier equipped with nuclear weapons for the purpose of sussing out military information. In both cases, the Department of National Security has opened an investigation to find the perpetrators and the main reasons for the attack. Beyond the suspicions and responsibilities that are yet to be verified, this event brings to the fore, albeit with a dramatic import, the deepest motivations for the design and publication of this book.

Philosophy still matters

This is not, however, an essay on computer security or a book prompted by a recent – albeit rather serious – chronicle event. Rather, it is the result of a long, philosophical journey matured over the last few years, aimed at highlighting the relevance of software code as the primary engine of our civilization, culture and contemporary economy. This code that takes multiple forms invisibly connects sensors, data, algorithms, machines, artificial intelligence (AI) and platforms. In this case, the chronicle has given visibility to something that is usually hidden from view and therefore underestimated. Usually we become aware of software only when it fails or when, as in other cases, it threatens our very existence both as individuals and as social beings. But software code is far more than what a computer war event could expound: it is a kind of "technological unconscious" that shapes and mobilizes our personal as well as our professional lives whether private or public. It is the invisible engine of our contemporary society and – this is the central point – it is also the condition that makes this world possible. This book was born, basically, with the observation that our culture lacks the speculative and philosophical view of contemporary and future software society. Its aim is to present and promote a more conceptual and speculative analysis on code culture. Time has come to think philosophically about software and its ecosystem: sensors, data, algorithms, machines, platforms.

Abroad, especially in the United States, there are presently in different universities curricula and courses (not to mention full programs and research projects) devoted to code. But the public debate on the cultural role of code, data, algorithms and artificial intelligence has only just begun to scratch the surface. In spite of its growing relevance, society is lagging behind in dealing seriously with the code revolution and its implications (consciously and critically – rather than relying on hearsay or taking matters on faith). But above all, in my opinion, we are culturally late in preparing present and future generations for digital thinking.

Today, however, *software takes command* – as Lev Manovich, the theorist of software culture, has recently written.[1] Contemporary society, economics, science and culture are strongly permeated and shaped by the software code embedded in processes, architectures, environments and objects, media and even implanted in humans, animals and plants. Faced with this pervasive and constituent presence, software as a central element of contemporary culture and society remains, however, still underestimated in its scope and its meaning, often limited and narrowed to issues of technological engineering.

In particular, this restriction tends to relegate software and coding practices to a subordinate and marginal position in cultural, epistemological, social and economic domains, not partaking into the discussions, reflections and analyses that take place in the more advanced research and development laboratories of the world. Meanwhile, *the code is eating the world* – to take a slightly apocalyptical tone. It is the software code that though hidden, activates or, as the case may be, deactivates transports and commercial logistics, financial transaction activities, digital multimedia productions, marketing and advertising automatisms, medical devices and tools, self-driving vehicles, new monetary technologies and so on. I am convinced that a philosophical lens can help to make this technological unconscious visible or, at least, observable.

This is a philosophically oriented book, but it is not a philosophical book. It is designed to arouse the interest of different readers: it is addressed to managers and business leaders, to public institutions and public figures, to the protagonists of social innovation and the third sector, to students who face technology from different disciplinary perspectives, curious about how computational evolution is morphing our world. It is not, therefore, directed primarily to philosophers. So, I beg them to forgive me if the language and the discourse are not academically disciplined. I remain firmly convinced that, avoiding certain language obscurities, the contribution of philosophical discourse to illuminate and stimulate a deeper technological understanding is more and more crucial.

I am also convinced, moreover, as sustained by Luciano Floridi philosopher of information,[2] that philosophy has to go back to dealing with the central issues of our present and near future, and not merely comment on and discuss the writings of past philosophers. However, the philosopher must have the desire and ability, strengthened by a conceptual heritage matured over time, to dig dirt again, to understand, to venture, to dissect and deconstruct the technological domains that have emerged in recent years. I will proceed in a lay manner on this path and through the various chapters, dialoguing with a selection of international philosophical texts that I consider relevant (largely unknown to the general public) to trigger and stimulate readers. As I myself have had the opportunity to do in researching digital philosophies for this volume. But not just in books.

Code, data and algos cultures

The writing of this book has benefited from a very special time and place: a long period of study, research and discussion at MIT as a visiting scientist. A summer and an autumn spent in an institution that has, in fact, been to me many places. First of all, the research center on complex sociotechnical systems, MIT SSRC (Sociotechnical Systems Research Center) inspired by professor Alex Pentland and recently affiliated to the new MIT IDSS Institute for Data, Systems and Society. But then, without any doubt, the "future factory" that give shape to this modern wonderland that is the Media Lab of MIT. Directed by Joi Ito, here innovation cultures pervade all disciplines from affective computing to civic journalism, from liquid interfaces to algorithmic cryptocurrency, from radical atoms to social physics. Finally, I have to mention CSAIL, the extraordinary center of research on computer science and artificial intelligence. All talented people that I am happy to thank collectively here for the stimulus and support they offered me. The responsibility for writing, of course, remains entirely my own. Special thanks, from everyone too, go to professor Alex Pentland, "Sandy" – as he is affectionately called at the Human Dynamics Lab – who welcomed, with openness and enthusiasm, the idea of this unusual, cultural and philosophical exploration. These thanks extend not only to the recognition of the personal and collective talent of this research group, but also and above all to the spirit of that profound vision for which innovation must serve to build a better world – as the MIT motto says.

The title *In Data Time and Tide* has, simultaneously a poetic value and a provocative intent. It is poetical insofar as it recalls an old English expression that today survives only in the proverb "Time and tide wait for no man". It is said to emphasize that people cannot stop the passing of time, and therefore should not delay doing things. In our case, it is inevitable to live in a world build by data, code and algos (I call it *a programmable future*) and we should no longer postpone to better and deeply understand our new world. But it also wants to be provocative. As we will discover at the end of this path, what we will talk about is nothing but a "proto-data" world. A world in which, due to the recent *deep learning revolution*,[3] data directly feed smart code, creating our ultimate world interface. Recently criticized by some as "the tiranny of data", it seems also a world largely closed and dominated by a few.

For this reason, to remain open and inclusive in the search for a positive and better construction, it is necessary to attempt a philosophically oriented exploration. Which is, by definition, the subject of this book.

Indeed, I would like this text to help widen our perspective on the possible directions that our future can take, a future where – as we will see in the different chapters – code, sensors, data, machines, algorithms and platforms can produce a better world. I have opened this preface mapping the destruction attempted upon our world by means of the code and data (or, better, by the humans that make use of them). In closing this introduction, I would like to propose, in the form of a wish, a different map of the potential creation of a better world. We intend to also underline the possibilities offered today by digital and AI (assuming, of course, that technology will be driven by far-sighted humanity). This map, of recent creation, shows the poorest areas of the planet discovered through the analysis of satellite images that are used as data sensors to which machine learning and AI algorithms have been applied. In this case, the intention of the Stanford researchers was to identify, with a data-driven and AI-driven system, the most disadvantaged areas where to engage supportive and recovery policies.

The design (not easy) of this new, better world, thanks to the help of sensing, mining and making technologies is, therefore, within our reach. Our explorative journey can now begin. With a last warning to readers. Maps of unexplored territories are built along adventurous paths, difficult and uncertain. We must balance, with confidence, courage and caution. By updating the motto of those learning to code, let's start by saying, "Hello, New World".

Finally, I would like to thank Egea and Bocconi University Press for having believed that the reading and reasoning that I was condensing might one day have the chance to become (also) an English book. It is my sincere appreciation for the fatigue, the intellect and the joint will of all people who, in various ways and at different times, have come into play in this adventure. And a special thank to Professor Derrick De Kerckhove for having appreciated the book and decided to make available his intelligence and knowledge in personally translating it. For me, it's an honor, a privilege and a real joy.

Notes

[1] *Software Takes Command* is the title of a recent book by Lev Manovich dedicated to the culture of code and software.

[2] As Luciano Floridi expressed in an interview with Sophia's Online Magazine: "Bad philosophy is disinterested in current problems. If you deal with what it means to make politics today in the information society, what is ethics in the era of *onlife*, what is knowledge in the era of big data and algorithms, where science is going when as today it is dominated by large research groups … it would be a philosophy of our time" (*The Philosophy of Information: Interview with Luciano Floridi*, www.lachiavedisophia. com – June 20, 2015).

[3] In *The Deep Learning Revolution*, one of its pioneers, Sejnowski explains how deep learning techniques (from language translation to driverless cars to voice assistants such as Siri and Alexa) is changing our lives and transforming every sector of the economy. Artificial neural networks can now play poker better than professional poker players and even beat Lee Sedol, the Korean Go champion of the world, as happened in 2017.

The Code

Software is Eating the Planet

The newly coded, encoded, supracoded, encrypted world speaks, of course, to the impossibility of ever understanding the code itself. The discussion of ontology has to be updated. The world is designed from macro to micro level to fuse the algorithmic with the ontological [...] So, what is this new ontology? M. Jarzombek, *Digital Stockholm Syndrome*

Software is now in command

Software code is nowadays increasingly embedded – installed within our world, could we say to stay on track – in forms and dynamics that are as obvious as they are, in reality, invisible. Software today embodies, to the utmost extent, the idea that the most influential technologies on human existence are those that, when they become familiar, disappear from view as such, becoming indistinguishable from life itself.[1] Who would think today about the book as a technology, for example?

We certainly talk of software in an indirect way, through its most obvious experiences: a playful application downloaded by millions of people like Pokémon GO, an on-demand sharing platform for rooms (Airbnb), a wearable bracelet for fitness (Fitbit), a self-driving car (my thought here runs on Google Car, Uber or Tesla), a digital factory with sensors, robots and AIs that Adidas imagined for its productive return in Germany, a new cryptocurrency (Bitcoin, for example) or legal contracts algorithmically executable on blockchain.[2] We could continue, of course, with many other examples. Or software is evoked when it is likely to fail or as it fails,[3] causing incidents and injuries, sometimes even lethal, to individuals or communities. The summer of 2016 provided us with two episodes of this paradox: while software was driving a Tesla car in an autopilot mode, it collided with a truck, causing the first death of a driver occupying an intelligent car; and again, software, on another self-driving Tesla car, drove another passenger with a heart attack while on the highway, saving his life. More recently, in March 2018, the first fatality of a pedestrian: a 49-year-old woman walking her bike across the street was killed by an autonomous car operated by Uber in Arizona.

Beyond these incidents, however, the software remains an invisible presence, pervasive and hidden at the same time. We are, in fact, in the same condition as Neo, the hero of the *Matrix* movie, before deciding to ingest the red pill that will allow him to see, with his eyes, the code that simulates the world. Like Neo, we are surrounded by code that, however, we do not see. Opaque is the software that animates our institutions, that organizes our cities, that mobilizes our lives, that keeps our money safe, that cares for our diseases, that matches successfully our dating preferences. Or, conversely, that endangers our own existence.

Many today agree that if electricity and the combustion engine have made industrial society possible, similarly, software is designing and building the new way of being in our world – the ontology, says Jarzombek, of our society and our future. It has also been noted recently that General Electric itself is turning into a software-driven company (as specialists say), having been for more than 120 years in industrial production areas such as civil aviation and military equipment, oil and gas extraction and transport facilities, systems for medical diagnostics and biopharmaceutical technologies. This change has been so sudden and accelerated that Jeffrey Immelt, former CEO of GE, in a public intervention observed: "If you went to bed last night as an industrial company, you're going to wake up today as a software and analytics company".[4] And GE needs to run. One fact for all: in the five years from 2011 to 2016, the top five companies in the world by market capitalization were no longer companies in the oil or banking sector, or low tech (Exxon, PetroChina, Shell, ICBC), but they were all powerful software companies (Apple, Alphabet, Microsoft, Amazon and Facebook), built on code that has taken on a peculiar ontological configuration as platforms.[5] And more recently (2017-2018) other tech companies are joining the $500 billion club: to name a few, Alibaba and Tencent. However, upon reflection, the comparison of the two rankings, spread over five years is impressive. Doubtless, we now live in a *code economy* – says Auerswald. An economy in which a new alphabet came into existence, an alphabet of only two letters, 0 and 1.

If it is true, as Lev Manovich suggests, that software has taken command in contemporary society, then new education and training initiatives, which are part of the US government program, are progressively spreading in various countries around the world. Programs like the one launched by former President Barack Obama (*Computer Science for All*) to include programming (coding) in schools testify to the importance of knowing how to program and write code for the future of the country. In his presentation speech, Obama was very explicit about the goals of this massive $4 billion education initiative:

> Now we have to make sure all our kids are equipped for the jobs of the future – which means not just being able to work with computers, but developing the analytical and coding skills to power our innovation economy. Today's auto mechanics aren't just sliding under cars to change the oil; they're working on machines that run on as many as million lines of code. That's 100 times more than the Space Shuttle. Nurses are analyzing data and managing electronic health records. Machinists are writing computer programs [...] In the new economy, computer science isn't an optional skill – it's a basic skill.[6]

Even the Chinese government, increasingly aware of what is at stake, has begun to push innovation by instituting programming classes for teenagers. To the point that the Tarena Learning Center in Beijing, one of China's leading Chinese language education institutions, has recently increased the number of enrolment from fifty to 4,000 students in just one year.

The relevance of programming is so high that some scholars are proposing computer science as a new, unexpected scientific domain,[7] the fourth after life sciences (biology and ecology), physics (physics, chemistry, geology) and sociology (sociology, psychology economy, history). Thus, appears a new disciplinary field to investigate that of computational sciences.

Technological unconscious

In spite of this growing recognition of the importance of the code, the latter still remains invisible. An invisibility that hides an unparalleled capacity for action, influence and government in the world. We are faced with a reality that, while active and economically, politically and socially relevant, is blurred or often absent from our cultural and societal reflection. Moreover, a philosophical understanding of the role of software as a key element and primary engine of contemporary civilization is today largely and dangerously deficient. I'm not talking, of course, about the technological understanding of the software code, that is, its operational, engineering or computational procedures. Typically, software is only thought of as instrumentality, as technology for producing services, applications, and platforms. Rather, I refer to the lack of philosophical analysis of that "invisibility" which is often cited, even in the introduction of software engineering manuals,[8] as a feature of the code and which is attributed to its "abstract" nature. As such – the experts say – software cannot be understood with our five senses. In fact, we cannot see, taste, tap, sniff or hear software. Of course, some tools can make some aspects sensitive: visualizing can help programmers to see code portions or a hearing aid can help you listen to the software while it is working. But – they reiterate – the underlying software belongs to "abstract objects" such as mathematical and philosophical concepts, and therefore, invisibility is part of its nature. Nothing strange on the horizon, it would seem.

And yet, I think that as philosophers we should attend to this opacity of software. I am convinced that the analysis of our new world and life will not be complete or even effective if we fail to make the software visible or, at least, make it observable even through the lens of philosophical speculation. Although it is used as an explanatory and enlightening concept for other phenomena (such as the hardware/software binomial to talk about body/mind or the concept of code to indicate the genetic program) itself remains an unknown object.[9]

We are faced with a powerful and paradoxical metaphor for everything we think is invisible and that, however, remains hidden, even while creating visible effects. This is what the philosopher Wendy Chun defines as the "invisible visible".[10]

Because of this opacity, software has also been compared to the kind of "technological unconscious" we have to deal with.[11] Hidden in the familiar everyday life both professional and personal of each person, this new unconscious remains out of view in at least four aspects.

First, it is still entirely under the dominion of scientific, engineering and information practices involved in programming, in the development of applications and architectures or, more abstractly, under the rule of the logical-mathematical disciplines. Therefore, those who do not belong or do not share these areas of knowledge are excluded.

Second, we are talking of an object that, as we shall see, is by its nature in a state of continuous evolution made of instability, adaptation, and discontinuity: updates, versions, upgrades, integrations constitute the new mode of programming software development, in an approach defined as *continuous design and delivery.*[12] Trying to understand a rapidly changing phenomenon can make everything much more obscure and impenetrable: what we understand today about the software may not be so illuminating tomorrow, and, more often than not, we cannot even read the software programs used in past.

Thirdly, being a commercial entity subject to industrial, market and intellectual property legislation, the code is protected from snooping by competitors or from malicious intents to exploit, reproduce or tamper with applications and systems. Corporations and governments often invoke intellectual property and security as reasons to conceal the code that is designed, marketed and introduced in the lives of consumers and citizens.

But the fourth – and perhaps more relevant – reason is that it is the code itself that is concealed, by its very nature, behind a complex and articulated series of processes, operations and languages that are, in most cases, objectively difficult to make visible even to those who write and read code by profession and training. In this case, we are confronted with a constitutive invisibility (difficult but not impossible to disclose) that produces other features, usually assigned to the software as effects. That too will have to undergo critical analysis: its immateriality and intangibility, its neutrality and ideology, its technical-engineering and above all its social nature.

In the face of this intangible and pervasive daily manifestation, an analysis and an understanding of what software "is", in my opinion, is more and more urgently needed for several reasons. As in studying and analyzing the industrial mechanization, we have understood the philosophy of the organization of factories, logistical movements and productive processes of industrial capitalism, similarly we should investigate and examine software (and the logic, dynamics and ecosystem that support it) to understand the philosophical structure of informational capitalism and its emerging forms: the platform business models such as those of Amazon and Uber; the immaterial work to which the social media urges us; the dominant logic of digital service or experience companies such as Apple;

the new industrial digital twins of GE; the artificial life of social chatbots and of intelligent agents in Facebook;[13] the token economy of cryptoassets, cryptocommodites and cryptocurrencies (Bitcoin or Ethereum blockchain and many other monetary technology innovations). These innovations will come to impact even more our economy and life than protocols on which we have built the web over the last twenty-five years. Incidentally, it is not surprising that the October 2017 issue of the *Metaphilosophy* journal was entirely devoted to the "philosophy of blockchain technology", as the second most relevant protocol layer on the internet after the web.[14]

For the philosophical analysis of software (as envisioned by coder and philosopher Federica Frabetti), however, it is not enough to be technically able to read it or to see what it is capable of producing, or to interpret the social and economic reasons for its development or even to identify its proper legal category. Studying software philosophically does not mean considering it merely as a cultural artifact, which can be explained away as a derivative of society and culture. Otherwise, its meaning would be derived not from itself, but by the context from which it emerges. We will not therefore engage in a socio-cultural or sociopolitical analysis of what software "makes" (engineering issue) or "what it wants" (political issue), although there are relevant investigations intertwined with ours – as we shall see – but rather a philosophical and ontological analysis of "what is it". We need to fully understand the meaning and nature of its being and, better still, its ontogenesis, that is, how software became what it is.

The first systematic multidisciplinary attempt to remove the veil of opacity and philosophically digest the sense of this technological unconscious comes today from so-called *software studies*. Starting with the insights of Lev Manovich, these studies gave rise to an international analytical (American above all) movement that began to put the code in the center of cultural, theoretical and speculative investigation.[15]

What does it mean – these studies ask – to live in a software society, to live in a world where software becomes the dimension that permeates all areas of contemporary life: communication, media, representation, simulation, decision, memory, vision, writing, interaction, and much more?

If Manovich's original question was what became of media after the advent of software ("media after software"), here we will have to try to expand the spectrum of analysis and ask what happens when the software becomes – philosophically speaking – the new human 'horizon of experience'. What we will have to do here is to apply the philosophical perspective of software studies to the industrial internet (industrial software), to the sensory environment (environmental software), to the quantified self (personal software), to marketing automation (marketing software), to chaotic logistics (logistical software), to decentralized organizations (organizational software), and so on. We have to ask not only what are media after software, but what is the "world after software", the world after the advent of code empowered by sensors and data, which incorporates algorithms increasingly driven by artificial intelligence and which is constantly instantiated today in powerful socio-economic platforms.

If – as we shall see – "code is modeling the future as the future is inscribed in the code",[16] then it is also about creating in our society more generally a sense of urgency for its philosophical understanding. This urgency is also vital in a narrower business perspective. If you want to innovatively develop new services, products and markets, a profound knowledge of the nature of software is crucial. And, in fact, philosophically, software deepens our concept of "what is possible". In short, the code redefines, ontogenetically, the new conditions of the world's possibilities.

This sense of urgency to interrogate is also growing with the awareness and apprehension that we are not and will not be able to understand this world we are building. The software-based socio-technologies we are developing are, in fact, a big part of the complexity we face. And although our ability to describe, understand, and act with complex systems is wide (thanks to individual genius and collective intelligence), as humans we have limited cognitive abilities – says Arbesman. The interconnection between systems, processes and applications creates complexities that neither our powers of abstraction nor our specialized skills are able to master. Internet of things, industry 4.0, smart homes, autonomous vehicles, artificial life, synthetic biology and nanotechnology, algorithmic finance and quantum computing will only increase this complexity. And, in fact, we realize that complex systems have a – powerful and even destructive – ecosystem dimension as they *emerge* (that is, in their ability to generate the unexpected). Suffice to recall the classical example of the Flash Crash episode, the collapse of the financial markets of May 6, 2010, triggered by abnormal transactions performed by high-frequency trading software.

To give a simple measure of the growth of this complexity in software – complexity that is commonly measured in the number of code lines that the program contains – the Photoshop source code lines (used for computer graphics) have grown forty times since 1990, while in the last ten years the Windows program has doubled. Today, for example, developing a mobile application requires forty thousand code strings, a car (even non-intelligent) contains between five and ten million lines, while the internet services offered by Google demand writing two billion lines of code. And it is not merely the internal complexity of each program that has grown. Because of interconnection and interoperability between companies, systems, and programs (for example, Google Maps and its Uber taxi fare display), this complexity has become extremely interdependent and humanly difficult, if not impossible, to handle. Nor can it be said that building this interoperability between systems is always done with full knowledge of what is being built. It follows that often the short–medium and long-term consequences of the code put into the world are often ignored and, in any case, difficult to anticipate.

Code meets philosophy

If, therefore, this is the time we are called to live and if philosophy is, as Hegel used to say, learning to grab our time with thought, then it is also time to think philosophically our coded world. In our case, philosophy is a bit like that red pill that at the end Neo, the aforementioned protagonist of Matrix, decided to take to finally "see" the code that creates his world.

It is time to broaden the study and analysis of software beyond the domains and disciplines of pertinence considered more natural and obvious (information, computer science, information engineering, logical and mathematical sciences). It's time to philosophically think about the software, to investigate it per se – as Federica Frabetti writes in her recent book *Software Theory*.[17] To do this, we must start with the classic philosophical question: what is software?

Beyond the easy technological descriptions, the answer to this question actually brings to light a quite mysterious object. A first dark matter that needs to be clarified is the semantic universe we use to describe this complex phenomenon. In fact, up to this point, we ourselves have blithely used a set of different concepts related to software technology. We have used indiscriminately terms such as "code", "language", "documentation", "program", "programming", "writing", "reading", "application", "operating system" various combinations of these as "programming language", "code writing", "software code" and so on.

To begin with, we have selected the term "software" as the barycentre of our argumentation. The word was supposedly coined at the beginning of the 1950s and began to circulate with an article by statistician John Wilhelm Tukey in the *American Mathematical Monthly* in 1958. It is opposed to "hardware" to point out, in systems of processing, the importance of logical-mathematical instructions as light or soft elements as compared to the heavy or hard components of electronic computers of the time. "Today, the "software" comprising the carefully planned interpretive routines, compilers, and other aspects of automative programming, is at least as important to the modern electronic calculator as its 'hardware' of tubes, transistors, wires, tapes and the like".[18]

The computer as a set of hardware components is therefore not able to work by itself. There need be programs as sequences of instructions executed to perform certain operations. And the set of all these programs – experts say – is named "software".

This conceptual separation was also enshrined in a historic industrial move in 1968, when IBM decided to divide the company between its software and hardware units. From that date, both symbolically and factually, software development and marketing became a new industry separate from that dedicated to the production and sale of hardware components. For *IEEE*, the reference scientific magazine, while "hardware is physical, software is intangible". Everything is clear and straightforward then? Well, maybe not.

Properly understood, software is defined today as a computer program written in a programming language and printed on accessible and legible paper documents (hence a document, a written text). But software is also at work when a computer executes a program (hence, an executable code), the present state of which is understood, for example, by looking at the screen. And it is at this point that the various components of the software are made visible, from time to time, as language, writing and code. It's not by accident. As we will see, software seems in fact to emerge as a material inscription operation at the intersection of language, writing and code.[19]

Engineers and computer scientists describe software development as an advanced "writing technique" that translates a text or group of texts written in natural language (the "specifications" of the software system) into a text or group of texts in binary language (called "executable code") that a machine can interpret and execute. Some distinguish between "code", understood as the textual and socially understandable writing, control and distribution practices of the source code, and "software" as the code compiled in an executable format including the final product in the form of operating systems, applications or functionalities. For our purpose, it is sufficient to say, in a nutshell, that nowadays a comprehensive definition of "software" should not only refer to software programs, but also include the writing and technical documentation relating to the writing of the programs themselves, in short, both *instruction* and *execution*.

Software, therefore, basically constitutes a "form of writing". When, referring to the code, we say "read" or "write" we are not using metaphors. We are actually describing the operating modes used by the program.

Note, however, that for computers, the writing operation is at the same time a read and write operation. Additionally, for humans reading and writing are not the same thing as they are for machines. In any case, for anything to be called *writing* requires necessarily a material inscription, a materiality that is a constituent condition of writing: "the condition for code to function is the possibility of inscription".[20] In this primary sense, the software is material. This is not only because it is created in historical conditions and particular materials, or because it is concrete in the effects it has on our lives, but because it is "written" and writing a language is a material operation.

The code is certainly a language, but a particular type of language. It has the characteristic of being "executable". Indeed, for some it is the only language that has this ability: "it cannot be said that the main purpose of software is recording in the same way that it is for writing or cinema. It could be argued that the main purpose of software is to make things happen in the world".[21]

While the natural language that we normally use produces changes in people's minds and behaviours (with uncertain outcome, I should add), the code, once prepared to function properly, is a language that produces exactly the effects that are entered in the commands. It is a language – the first and only one – that "does what it says". The code would have no other reason than to give instructions to a machine. It is therefore a writing that calculates and commands.

Code is a language that is written and read, though not in the sense that we are used to reading and writing. It must be written and read in a form that a machine can understand and activate. And – Frabetti adds – the machine, in effect, is the ultimate judge of the fact that this writing can effectively act in the world.

The code is "action" in a twofold way: events happen in a computational machine – it changes its status and behaviour – and, as a result, things happen in the world. And this is a key point in our argument that will become even clearer in the following chapters. In this sense, programming is a highly performative and future-oriented writing (that is, at some point, part of the history of software engineering). The code is an instrument that is not primarily used to record, store, or remember as it is for other types of "writing" such as literature, television or cinema. It is not conceived in terms of what happened, but it is projected to produce events about to happen. It is a form of material writing of the world: from its unfolding, literally but concretely, the world emerges. So, it is the future, not the past, that is inscribed in the code.

It is important here to point out the evolution of the software development paradigms (without telling the story in detail) to integrate some of the fundamental elements that define the current ontology of a world that, being written by the code, becomes "programmable."

The history of software has known three great paradigmatic ages: from the handcrafted beginnings to the standardized discipline phase to the agile approach of our day. It is a story that, in the early 1950s to the mid-1960s, saw programmers, electronic engineers and mathematicians develop relatively simple software applications craftily and ad hoc, almost artistically and without specialist knowledge or vertical skills; a situation that continued until the mid-1960s, when the growing success of software applications and the need for ever more complex architectures led to a crisis in the craftsmanship of the beginnings. By the end of the 1960s, during the time when hardware and software were conceptually and industrially separated, began a period when the software developer community engaged in a deep criticism of the nature of software, of its development models and its technological, industrial and social criticalities. The Garmisch Conference of 1969 initiated what is known as the "software crisis": from that discussion, the community of programmers will be coming of age – writes Frabetti – with a key identity step. The "problem" software (the growing complexity of development, the risky impact on society, the industrialization of programming) becomes the "solution": the tool to solve specific issues or needs through analysis, design, implementation, and to verify solutions.[22]

In the early 1970s, this second historical phase marks the birth of the software engineering discipline and is defined by the so-called "waterfall" development approach: starting from the breakdown of the problem, the needs of users and the technical requirements analysis, the developer will describe and design in detail the specifications, modules, interactions, algorithms and the data structure that the software system will need to have (*designing*); then programmers will write the code that the system will have to propose a solution (*coding*) and finally test it and validate it (*testing*). It should be noted that during these phases, the job is to produce the project documentation that will guide programming the code. And, in fact, the separation between design and production is more of a division of labour than a substantial difference.[23] The name "waterfall" designates a process that is governed by the requirements of the solution and until recently it has been considered as the standard approach to software development.

However, in recent years, this paradigm has been subjected to criticism in particular because of the so-called "volatility of requirements:" during the development phase, a constantly evolving world (regarding different dimensions: markets, regulators, consumers, competitors, technologies) increases the rate of requests for changes and upgrades in software requirements, rendering the process incompatible with the waterfall approach that is based instead on the immutable pluri-annual planning of initial requirements.

On the contrary, today, development by iteration – called the "agile" method – supports writing code so as to remain open to requests for changes suggested by the client. An approach is thus adopted for short cycles of analysis, design, production and verification, which are constantly being repeated in order to include modifications, adjustments and revisions of the software during development. This new technological development paradigm has very strong philosophical implications. It is in fact able to support the ontology of an updateable world with continuous design and delivery or, from another perspective, called *devops* (that is, development and operations gradually becoming almost inseparable).[24]

If we want to translate this philosophically, it means that in the new digital ontology we have become creatures of the update: "To be is to be updated: to update and to be subjected to the update," as Wendy Chun says.[25] The mode of existence of digital objects (mobile applications, data platforms, smart objects, crypto currencies) is to be consistently subject to updates, whether such are deliberate or simply routine, whether correcting malfunction or addressing non-functioning programming. "At the beginning, there was the Word; now there is the Update",[26] says Jarzombek provocatively. In this new world, the failure to update can also cause profound legal and criminal repercussions. For example, in the case of a self-driving car accident, the owner, in spite of not driving it, is nevertheless considered guilty by American and English law if he or she has not kept the software installed on the machine up-to-date. In the future, for companies marketing autonomous vehicles, the responsibility for an accident will be the onus of the car makers (no longer the non-driver), but the failure to update will still put the owner in the condition of being punishable and prosecutable by law. The failure to update not only puts at risk operating a service, product or platform and therefore its being in the world technically, but also jeopardizes the legal (and penal) integrity of actors involved in an autonomous ecosystem, including reciprocal relationships and interactions.

A second important aspect implied by this updating ontology is on the side of training and education, human learning and understanding of new technologies. Given the ever-closer cycles of acceleration, obsolescence, and technological innovation, the ability to quickly forget and re-learn rapidly from consumers and employees will be crucial. In fact, many of the applications we would normally interact with in the future have not yet been invented, and those that are emerging will come through rapid procurement and deployment processes, while those on the market will be constantly upgraded to enable new features and interfaces. In the face of this state of continuous change, knowing how to "unlearn" will be a professional and personal competence, at least as much as knowing how to learn quickly or to be able to search for knowledge and information online.

Hidden and sudden

We still do not have the ability of Neo to see the code running (at least not in its simplicity and immediacy), but we are beginning to understand its relevance and to reveal the opacity. We now have to go back to the code itself and continue, going a bit deeper. In order to be executed, to become operational and to operate in the world, the program written in a high-level programming language (*source* language) must be reinscribed in a machine-readable binary code language (*target* language).

You cannot run the source code: it must be interpreted and compiled appropriately. And this is another key step.

This automated reinscription process is defined as "compilation" and the compiler is a program which, by the way, reads and translates (to use a metaphor) the source language in target language thanks to a reinscriptive process. As Frabetti clarifies, the compilation process has basically two phases: in the first one (*lexical analysis/scanning*), the compiler during the translation checks the correctness of the syntax of the source language and divides it into subgroups called lexemes, that is, the strings of code in relation to specific separations in the syntax. The lexemes produced by the lexical analyzer are in turn transmitted to tokens that completely rewrite code strings with the addition of explanatory symbols.

In the second phase (*syntax analysis/parsing*), the parser checks and re-writes the structure of the language, starting from token, to generate an intermediate code form (conceptually similar to a tree representation). In fact, between the source code and the target code there are a number of intermittent rewritings that are called *pseudo-code*.

It is important to emphasize here that in these rewriting phases the compiler can determine and recognize any error in accordance with the lexical and syntax rules but – and this is a key point – it cannot be ascer-tained that the source code will work in the expected way or if unexpected behaviors will arise. The translation or interpretation of source code in executable code is not an isomorphic operation, but is more of a logical equivalence operation. This implies a certain destructive and creative skipping, at the same time, of the translation during which the ultimate meaning and actual code action coexist. Each rewriting, potentially and in fact, can change the original meaning of the code and its future potential action on the world. And this clarifies yet another level of opacity in the constituent dimensions of the code by introducing us ultimately to the issue of its "fallibility" – which we will see later.

Continuing in the path from the source code to its final translation in circuits, the intermediate code is rewritten by the generator code, the last stage of the compilation process, in the target language. The generator code, starting from the intermediate code, maps the target code: if this is a machine code, logs or memory locations are selected for each variable used by the program. Intermediate instructions are translated into machine in-struction sequences that will be called upon to perform the specific task. The code becomes executable and thus executed. It now has the power to operate in the world as well and even more so when compared to other languages and scriptures that aspire to sovereignty and dominance,[27] as much even as the power of social discourse, for example, legal, political or medical.

This synthetic technical description of the code writing process allows us to begin to extirpate the software from invisibility.[28] And it makes us understand better some central dimensions and perspectives for the future of the development of digital economies.

Meanwhile, the first thing to observe is that the distinction between hardware and software is fragile and unstable. As Frabetti points out, although commonly accepted in engineering and computer environments, von Neumann's distinction at the end of the 1960s is more related to the division of labour (between the design of the software and its coding) than to the ontological issues: philosophically rethinking the code means then to also understand how this boundary, being neither given nor natural, can easily be traversed, thus prompting us to imagine – as it happens – a hardware that is not traditional (bio is the new hardware) or interfaces that do not have the possibility of appearance (the best interface is no interface) or, again, an infrastructure treated as code and data (infrastructure as code).

The distinction is therefore not only less clear and immediate than one would imagine, but is also progressively eroded. And, paradoxically, Ashford Lee wrote that hardware is ephemeral, software endures. We go to an increasingly "soft" hardware and vice versa, we develop more and more software to hardware, to the point that it will be more and more complex in the future to decide where one ends and where the other begins. If we mean by *hardware* the kind of physical support (with its circuits and microprocessors, tangible also through its peripherals such as keyboards, mouse, etc.) and by *software* (assumed as intangible) a sequence of instructions given in the form of programs, in reality this distinction "emerges from the process of internal differentiation of software as material inscription".[29] It is not given by nature, but is negotiated in the generating process of the code of instruction, inscription and execution.

We will return to this aspect in the final chapter when we talk about the governance of platforms and the "stacked" world, but it is useful to anticipate that all the software companies we have quoted in the capitalization list are moving rapidly to the conquest of even the hardware. From the famous "Area 404" of Facebook (presented to the public in August 2016), a factory where the social network will build the prototypes of its solar drones and its visors for virtual reality and much more, to Google that has decided to produce at home (and no longer outsourcing as it did for the previous devices) Pixel, its latest smartphone launched at the end of 2016. Not to mention Google Home (released globally in 2017) or Google Car, which represents the future of the crossing between the so-called *new hardware* and the internet of things. On the other hand, to thrive in the automotive industry (if you can still call it so), the construction of the intelligent car will in all likelihood require the emergence of partnerships – the first signs of it – between automotive makers (hardware) and software companies (such as Google and Ford, Volvo and Microsoft, General Motors and Lyft). Another recent example: according to Bloomberg magazine, in April 2018, Facebook is assembling a team to build its own chips adding to a trend among tech companies to self-supply hardware lowering their dependence on chipmakers. In the past, Apple also started shipping its own chips across many of its product lines, while Alphabet has developed its own special artificial intelligence chip as well.

A second major philosophical aspect is the emergence of the unexpected or the implicit fallibility of the code. We know, at this point, that compilation is the process by which a program is executed through its translation from a high-level code to programming the machine language. But, as we anticipated, this translation does not guarantee that unexpected, potentially destructive events will not occur during its operation. The software, in fact, lives in a paradox: innovation anomalies must exist so that the software exists, but they must be eliminated so that the software becomes stable. And thus, malfunction is a key time to disclose the nature of the code because "the circumstances in which software does not work – better, in which software does not work as expected –, could tell us more about software then those in which software works".[30] In addition, it is when a software development malfunction occurs that is the critical phase during which to make a decision about whether it is actually a dysfunction to be corrected or an anomaly to develop and integrate as a new feature inside the system. There is a famous saying among experienced coders: "Any sufficiently advanced bug is indistinguishable from a feature". And it is when an anomaly emerges – and it will assuredly emerge – that it will become necessary to make a decision, and that could be not only a technical decision, but a strategic (and political) one in the broad sense, concludes Frabetti.

Precisely because the software may fail (and the probability of its failure should be considered unavoidable), then it becomes clear as to how to change, for example, an operational approach to manage the failure in development. If failure is intrinsic to coding, you will have to abandon the traditional philosophical approach that maximizes the time between this failure and the next one (MTBF, *mean time between failures*), that is, to do everything in order for service not to fail and to focus on a paradigm that tends to minimize the time for the repair of the service failure (MTTR, *mean time to repair*) and activate the mechanisms and procedures for timely resolution of the fault.

The software, therefore, works as a form of material inscription that invisibly (and with a development not always clear) is called to make things happen. This brings us to the third philosophical aspect: the question of software's ability to act in the world (*agency*). At the beginning of 2016, the US National Highway and Transportation Security Administration stated that in the case of self-driving Google Cars it is the driver, whether actively driving or not, that is to be held responsible according to federal law. But what does it mean to have the ability to act in the world? A non-intelligent car, on average, already contains between five to ten million lines of code. But, in reality, it is an insufficient and "unintelligent" code. For a car to be autonomous, it must be able to sense and understand the outside world and not merely execute its own operation. A self-driving car must have sensors and actuators – as we will see in coming chapters – and above all, it must have a smart code to be able not only to calculate but also to judge and act accordingly. Attributing code to feel and act on the world (*agency*) is, in any case, absolutely essential. It is the precondition of any social and real world relationship granted by the code.

To accomplish this, software has had to change its deep nature over time. A crucial historical moment in this ontological evolution was the transition from *batch* code processing to *interrupt*. At the beginning of the software story when launching a program, it was necessary to wait for the end of its execution to be able to switch to other operations or change the procedure (*batch processing*). Switching to the interrupt paradigm (with the possibility of interrupting the program due to external events) has removed this technological limit, which is also a cognitive and operational limit for the code to act in the world. Philosophically, interrupt is the state of the computational machine open and ready to be disturbed by the interference of external events, typically represented by peripherals that call back and require attention.

As Yuill wrote:

The interrupt fundamentally changed the nature of computer operation, and therefore also the nature of the software that runs on it. The interrupt not only creates a break in the temporal step-by-step processing of an algorithm, but also creates an opening in its "operational space". It breaks the solipsism of the computer as a Turing Machine, enabling the outside world to "touch" and engage with an algorithm. The interrupt acknowledges that software is not sufficient unto itself, but must include actions outside of its coded instructions. In a very basic sense, it makes software "social," making its performance dependent upon associations with "others" – processes and performances elsewhere. These may be human users, other pieces of software, or numerous forms of phenomena traced by physical sensors such as weather monitors and security alarms. The interrupt connects the dataspace of software to the sensorium of the world.[31]

In some contexts, the software acknowledges full agency (for example, in the case of some AI programs). It is more often believed that it possesses a "secondary agency",[32] that is, having the capacity and the ability to support the agency of any other primary agent, that is, more often than not, a human being (a user, a programmer, a company, a government). Whether it is the presence of software agents in the form of small pieces of code that are delegated to specific tasks or the deployment of models of wider intelligent software architectures, for us it is important to understand here how the code is saturated with agency (*agency-saturated*).

Even when it is invisible or hidden within another technology, or the environment (a car, a home, a chain of furniture, an office, a body, an accessory), the behavior we expect from the code is to be able to act in its own domain or in the context in which it is embedded. That this agency be distributed among human beings, environments and things has long been a recognized perspective in disciplines such as social studies in technology or the sciences of cognitive anthropology. But even after considering the attribution, delegation, and distribution of agency associated with code, we still have some difficulty in saying *what it is the code does*. Is it not a fact, asks Mackenzie[33] that the code is "a direct expression of human agency in relation to things like *start, move, stop?*"[33] However, deciding where to position the agency of the code, or how the code distributes the agency, and who or what possesses the ability to act becomes in many cases rather complicated. It is increasingly difficult to specify who the critical or primary agent is in any given situation: is the agency that of the programmer who wrote the code that contains certain instructions? Or, vice versa, is the program acting on behalf of a user as a third agent? Or still, is it a matter of having the software operating under certain constraints and rights? But in an environment populated by bodies, things, systems, and smart processes interacting in multiple scales and with unexpected dynamics – topics that we will examine in detail in the chapter on artificial intelligence and experience – how to recognize and assign this new distributed subjectivity?

A programmable world

Lastly, we have to face a more general point: why do we say that *software is eating the world*? This is the expression coined by venture capitalist Marc Andreessen that has been circulating for some time among tech events, business meetings and academic conferences. In an article written in August 2011 for the *Wall Street Journal*, Andreessen, the creator of the first browser (Mosaic), explains in detail why companies such as Amazon, Netflix, Spotify – software companies – were eroding margin and value of the not-yet-software-driven competitors (Borders, Blockbuster and traditional companies).

The phrase expresses, with alarming and apocalyptic tones, an idea increasingly shared between analysts and futurologists, startups and venture capitalists, business leaders, managers and professionals. This is simply to mean that software code is developed, distributed and activated pervasively and continuously to metaphorically and even factually, attacking and destroying the world as we have known it so far (and, with it, the society, the economy and the way in which we produce value). This happens through mobile and/or wearable devices or appliances installed within industrial and market processes or, moreover, embedded in home and industrial environments, urban and domestic spaces, new monetary technologies and platforms. Indeed, in short, we are witnessing the destructive end of one world and the emerging of a new one.

Of this destruction, widely documented by the economic and social chronicles that tell of technological accelerations, end of business-as-usual patterns, of inflationary media emerging thanks to artificial intelligence, software undoubtedly represents the key element, to the point of making it a phagocytic engine.

In this vision, the code is eroding, progressively and deeply, traditional patterns and practices of relationships between markets, architectures and production dynamics and the exchange of services, modes and purposes in co-creation of value. What we see are the impacts of innovation on business practices, on experiences that can be imagined and offered to consumers, on organizational forms and emerging labours, on the provision of digital services such as new forms of logistics of "constant contact" and "real time" (take Amazon, for instance).

Although from a different perspective (that of enthusiastic supporters of the new digital world), Andreessen's phrase reminds us of the tradition of dystopian narratives around the dark and destructive underpinnings of the code, characteristic of many examples in literature, cinematography, and sociology past and present. From the Cold War, in fact, our cultural narrative about digital technologies directed at/by software has privileged a dichotomous and mutually exclusive view of world and code: on one hand, an innocent and pure nature, on the other, an autonomous and malevolent digital technology. The first is material and positive, the second producer of hazardous waste and dematerialization. In this perspective, nature would mean to live fully in the world, while codified intelligence would lead, on the contrary, to the destruction of this natural world or, at least, to an escape from it. At any rate, computational technologies would distance us from the real world, distract us from our physical bodies, and erode our connection with space and the natural environment.

In this genre of apocalyptic narratives (popular examples are films such as *Blade Runner*, *The Matrix*, or *Terminator*), intelligent computing machines would bring humanity to slavery, would push consuming, leading to the destruction of environmental resources, destruction that would be compensated by the creation of a simulated and hallucinated reality (the fake world that the Matrix creates for Neo). Of reality, nothing would be left except the desert (to quote Baudrillard), a devastated and nightmare world, distrustful – when not disturbed and in conflict – of the encounter with technology and the code, in particular. Andreessen's phrase, of course, does not have this negative connotation: his direct recall is addressed more to Schumpeter's theory of creative destruction than to the exterminating force of Schwarzenegger. And yet this expression, with its apocalyptic formulation, prevents us from seeing the software for what it is and, consequently, its relationship with reality and with the world as we have, however, described it until here.

For most of the world's population, life, work, citizenship, sociality are all dimensions that today require the presence and activation of software programs. The code manages the efficient flow of fuel in our car as well as the energy consumptions of our appliances. We must therefore reject the current dominant narrative that the code is transcendent, immaterial, abstract and destructive. Likewise, we have to think differently about technology. We must recognize that technology is the *natural* way for humanity to be in the world. Man has always been technological, human emerges and evolves with technicity and with the technologies that it creates and puts in place. Ultimately, there is no way of distinguishing technology from culture and technology from humanity. We must admit that code is continuous with our world. What we must understand – as we will try to do in our philosophical journey – is what kind of world and humanity is emerging through the code. As anthropologist Paul Dourish wrote: "[Software] is philosophical in the way it represents the world, in the way it creates and manipulates models of reality, of people, of action. Every piece of software reflects an uncountable number of philosophical commitments and perspectives without which it could never be created".[34]

We will then have to learn to favor alternative narratives that tell how the code relates to natural spaces by privileging a new organism, a new holistic vision and an active interconnection between human and non-human agents. There are lateral and minority narratives in the history of the relationship between code and the world, but they have been re-emerging strongly in recent years. Not only is the code not immaterial but it is deeply rooted in matter. It is material for being a signs system, as well as a series of electronic impulses, a catalyst for physical state changes within a machine and what the machine controls, a compliant product in the form of software and a mediator between human intent and machine action – scholars say. In art, most recently, these instances of the code have emerged from the darkness of the source files, making them visible on screen and beyond.

There is talk about *expressive code*,[35] exhibited in various performative digital works of art and thus made sensitive and material. With these new aesthetic practices, we can better appreciate how the code is co-existent with the world in its various components, organic and inorganic, in spite of its invisibility (which, as Federica Frabetti reminds us, is not a precondition for its existence but which, at the same time, is more than a mere convention).

The code thus has its articulated and complex material reality: we have learned that software is an object that must be observed in all the different forms that it actually takes, from document specifications, to the source code and to the executed code. And indeed, software can be investigated in relation to the type of code we are observing (Assembler, C ++, Pascal), based on its status (sourced, compiled or disassembled) for its location (embedded, system or application), in the forms of representation (text, visual, or mapped as a graph) or depending on the phase of its life cycle or its political (free, open, or commercial) connotation.

We understand it to be on many levels. Code is distributed and embedded in many complex sociotechnical infrastructures; it builds on and works on high-level language and interface elements; it is integrated into the professional practices and dynamics of work and communication; it is operative at a micro-interaction level with the physical properties of conductive and non-conductive materials. So, we begin to understand how and why the ontology of computation (the analysis of the "programmability of the world") is becoming increasingly relevant for building our culture, society, economy and business.

We hear more and more talk about *programmable city*, *programmable money*, *programmable matter*, *programmable contract*, *programmable law*, *programmable life*. For this reason, software can no longer be thought of as an instrument, but will have to be re-thought as constituting the human. A long, inevitable path awaits us in the coming time and tide to recognize and value this centrality of software in the ways it exists and where we experience it. Finally, as we shall see in the following chapters, in the ways in which it produces the very conditions of possibility and existence of our world.

Notes

[1] The work of Mark Weiser, father of ubiquitous computing, is at the origin of the idea of "calm technology". According to Weiser, the success of the design of a technology is linked to its ability to become invisible when it enters and takes part in the constitutive and almost natural way of life without being invasive or requiring the attention of the user.

[2] The blockchain protocol enables a distributed, encrypted and public register that can support virtual coins, but also smart contracts for algorithmic execution and coding of contractual documents (contracts that are translated into executable code) and other digital services. We will talk more in detail about this in The Platform chapter.

[3] One example will stand for all, the fear of the so-called "millennium bug" with the arrival of the year 2000 and the risk of confusing software systems because of the two ending zeros, potentially confusing it with other similar dates (in annotations of years with only two final digits).

[4] Interview released during the Developer Conference organized by GE Minds & Machines (2016).

[5] To the subject of platform economy and platform thinking I will devote a specific ending chapter focussing on platform model. Here it suffices to argue that many analysts and researchers are now oriented to consider our era as the era of platforms, and to come to the hypothesis of a new platform capitalism.

[6] Extracted from Barack Obama's speech transcribed and stored on the White House website. Someone else in the software industry went ahead and re-launched: "Software is a mindset, not a skillset" (Jeff Lawson, founder of Twilio, at the Amazon conference for developers in December 2016).

[7] This perspective is explicitly stated in Rosenbloom (2012).

[8] See for example the introduction to the classic Rajlich (2012).

[9] Chun (2011).

[10] *Ibidem.*

[11] Clough quoted in Fuller (2008).

[12] A continuous design and delivery approach means that software development is an infinite process of designing and deploying constantly updated services and applications. See, for example, Sussna (2015).

[13] Chatbots are software programs that use artificial intelligence to simulate an intelligent agent or assistant in a conversation with a human user.

[14] In computer science, a protocol is the set of rules formally described and defined in order to favour and manage communication, flux, relationships and interaction between one or more entities within networks.

[15] "As software becomes a putatively mature part of societal formations (or at least enters a phase where, in the global north, generations are now born into it as an infrastructural element of daily life), we need to gather and make palpable a range of associations and interpretations of software to be understood and experimented with. While applied computer science and related disciplines such as those working on computer-human interface have now accreted around half a century of work on this domain, software is often a blind spot in the wider, broadly cultural theorization and study of computational and networked digital media" (Fuller, 2010, p. 3).

[16] See Berardi (Bifo) in the preface to Cox (2013).

[17] Frabetti (2015) is a key text for the philosophical analysis of software. The essay clarifies in detail and with great depth the aspects that we synthesize and simplify in these paragraphs. We point out in her text observations to support some of the concepts presented in this chapter, investigated in a tight and critical dialogue with those who abroad, through a speculative lens, are philosophically analyzing the software code.

[18] Quoted in Fuller (2008).

[19] This is the argument put forward by Frabetti (2015) who discusses it with the theory of software produced by the most recent philosophical thought.

[20] Frabetti (2015), p. 57.

[21] *Ibidem*, p. 17.

[22] This approach to coding as problem-solving is certainly the dominant one today. However, it should be noted that it is not the only one possible. Another approach considers and uses code above all as a tool for the expression, growth and enhancement of human creativity. This is, for example, the philosophical and educational perspective brought by the Lifelong Kindergarten of the Media Lab and its creative and playful learning experiences and projects. In particular, the group promotes creative learning through tools such as Scratch, a programming language and an online

community where toddlers, boys and girls can express their ideas by developing their own voice and identity through project creation and sharing such as art animations, interactive stories or videogames. I thank Carmelo Presicce, a researcher at the MIT Media Lab, for the discussions we had on this vision of programming as a tool for building worlds and not just for solving problems.

[23] Frabetti (2015).

[24] DevOps is a software development approach (but it is, above all, implicitly a business development perspective) in which the time of developing the code and the time of its release into production are very close. New micro-versions of the software are constantly, quickly, automatically developed and brought into operation with multiple releases throughout the day. Other elements of this ontology of the code translate into business operating processes: testing in production (test while living, not just while it is developing) and the operations to consider design as input (not just as output).

[25] In Chun (2016), we find this perspective on the ontology of the update, which, as we shall see, also comes back to other authors (Kelly, Jarzombek) to mean the centrality of this technological trend.

[26] Jarzombek (2015), p. 50.

[27] The term *governmentality* is central to the theoretic discourses of philosopher Michel Foucault. It purports to meant that specific art of governing that, through a set of institutions, procedures, apparatus, analysis, reflections, calculations, devices and tactics, assumes taking over the population and guaranteeing government for the living. By *government*, the French philosopher refers instead to the capacity to structure the scope of action of others and thus to govern their behavior.

[28] For a more detailed description, see Frabetti (2015).

[29] Frabetti (2015), p. 144.

[30] Ibidem. This political awareness has been explicated more recently even within the historical developer community. "Civilization depends upon us. In ways it doesn't yet understand. In ways we don't yet understand" according to Robert C. Martin – among the signatories of the Agile Manifesto – in the video *The Future of Programming* in May 2016. In Martin's talk, this awareness is accompanied by an ethical warning. Since the software's failure (endo- or exogenous) is inevitable ("a significant tragedy will happen – says Robert – and then they will accuse us of being responsible"), Martin urges the community of coders to create a deontological discipline (as for doctors and lawyers) who self-rule software development before (this is the fear) others can impose, from outside and inexperienced, a regulatory policy to software.

[31] Yuill (2008), p. 162.

[32] "Secondary agency", says Mackenzie (2006).

[33] Mackenzie (2006).

[34] Quoted in Kitchin, Dodge (2011).

[35] Swanstrom (2016).

The Sensor

Senses, Sensorium and New Sense

This is the becoming environmental of computation [...] If the satellite view has largely been narrated as a project of seeing the earth as a whole object, then the more distributed monitoring performed by environmental sensors points to the ways in which the earth might be rendered not as one world, but as many
J. Gabrys, *Program Earth*

The empire of sensors

All living things perceive, process and act in the world in different ways. Our senses are able to capture and compute a significant amount of stimuli and information we are exposed to, allowing us to interact with the environment that hosts us. As human beings, we can "read/write" – as neuroscientists say[1] – via signals to and from the central and peripheral nervous systems, and remodel and adjust our behavior, our growth and our survival as individuals and as a species. Thanks to our five senses and to all of our cognitive abilities, we live in the world together with creatures that also possess special sensory capabilities.

But why only five? "You have five senses, I have six" says Forrest – whose identity we will not reveal because he has asked to remain inconspicuous – smiling and showing his arm with the invisible magnet that has been implanted recently under his skin.[2] This contraption serves to expand his perceptual capability by altering his body. With this magnet, it is possible to overcome the limits of our biological hardware. As an expert biohacker, Forrest added a sixth sense: the ability to perceive magnetic fields. It is a unique – and you might say extreme – case of someone who is driven by the desire to try a sensory apparatus increased technologically. It is a niche market surely, but not the only one. In any case, yes, we will all continue to have five, more limited, senses. That is, perhaps, only for a while.

Human eyes perceive only part of the electromagnetic spectrum, wavelengths between four hundred and seven hundred nanometers, but we cannot perceive ultraviolet, infrared, gamma and X-rays. The same is true of our hearing that can only perceive frequencies in the Hertzian acoustic range from twenty to 20,000 cycles. Other animals do not have these limits. Bees can see the UV rays, thanks to which they find flowers, while bats, dogs, dolphins, insects can perceive the ultrasound spectrum. In this, they have the same senses as our own, yet on different scales and sizes. But they are also enriched with non-human sensory equipment: for example, echolocation or biological sonar, that is, the ability to perceive distances, environments and objects by estimating the returning echo of the sounds they produce. Similarly, sharks and stingrays that, thanks to specific internal organs (conductive gel canals that can perceive an electric voltage change), can use sensitivity to electricity to locate other animals in low light areas.

Fish, spiders and many other insects can feel the polarization of light because they are equipped with photoreceptors along with detectors, chemo-receptors, thermo-receptors and so on; some birds (and mammals) can hear the strength and direction of the magnetic field through specific magneto-receptors that they use to navigate in flight. Animals therefore modulate senses like our own according to multiple scales, but they also reach levels of sensitivity that humans do not possess. In each organism, evolutionary pressure has led the body to prioritize certain sensing skills over others (and hence to elaborate the information needed to act in their environment). The way we perceive the world is therefore based – sticking to the metaphor – on our current hardware and software. But all this is rapidly changing, albeit often, inadvertently.

In our attempt to better access the world (or many different worlds, as Gabrys suggests in her book *Program Earth*) we are developing a new sensory apparatus made of software code, devices and mobile accessories, environmental sensors, data and artificial intelligence incorporated into our daily and professional life, in, on, or near our bodies, carried by our movements inside our homes as much as in the environment, in modes, scales and shapes that have no precedent in the history of humanity.

We are not aware of it, but a new sensorium – a new ability to listen to the world through data – is emerging. We are all becoming a bit like Forrest, albeit with less surgical and invasive technologies than his. We are all, of course, soft biohackers.

In 2020, to give us new ways of perceiving the reality around us there will be 34 billion devices to access the Internet's content;[3] of these, 24 billions will be new, the other 10 billion consisting of existing intelligent devices such as smartphones, tablets, or watches. The sensor market is witnessing in fact the strongest expansion in all industrial sectors and in all markets with an upward trend throughout the coming decade. This growth is spearheaded by the USA and Europe up to 60 percent of the market share, keeping in mind that the expanding Asian growth is projected to advance even more rapidly. It is an industrial and commercial development driven mainly by the demand for industrial sensors and, in particular, for sensors based on micro-electromechanical systems for the next generation of intelligent automotive industry. But the demand for new consumer electronics (mobile) and intelligent objects also invites significant initiatives, along with the demand for sensing – linked to smart cities – for health and safety, energy and services, building and home automation, transport and retail.

For example, an average American car is equipped with about seventy sensors, including gyroscopes, accelerometers, ambient light detectors and humidity diffusers. Cars are not new to sensing.[4] However, those built before the year 2000 have only a few tracking devices (anti-theft control, airbag safety systems, traction control and speed, acceleration and tire position). It is only after 2000 that assisted or totally automated driving required the installation of new and more sophisticated sensors such as those meant for self-parking or lane changing on highways. They allow a car not only to perceive itself, but also to evaluate its position, to artificially recognize its surroundings, to anticipate what is going to happen on the street and to plan its response. This involves artificial intelligence for sure, but also artificial perception. The arrival of *machine* learning and *deep* learning – as we will see in the chapter devoted to artificial intelligence – has been a very strong motivation to develop intelligence in cars that, through sensors and actuators (the mechanisms that enable steering, braking, speeding in such a way that a car perceives and understands the world) are reaching environmental autonomy. For another piece of evidence: in the first trimester of 2016, sales of car-to-internet connections rose by 32 percent in the USA and exceeded those destined to smartphones (31 percent), tablets

(23 percent) and internet (14 percent). In the Usa (the largest revenue market), the connected car penetration is currently at 38.8 percent (Q1/2018) and is expected to hit 81.6 percent in 2022.[5]

Turning to a second example, take the sensors that today equip drones, small aircraft, remote or autonomous land or sea vehicles.[6] Conceived to protect human activity, with remote control in extreme, hostile or at least risky environments such as catastrophic situations, impervious locations, or contaminated areas (say, mining jobs, firefighting, physicists in radioactive laboratories, astronauts in space or those who operate in oceanic waters), *drones* are increasingly employed in areas such as construction and precision agriculture, industrial internet, transport and logistics, security as well as – undoubtedly – in the war industry. To cover the entire electromagnetic spectrum – from gamma rays to X-rays, infrared and ultraviolet, from microwave to radio waves – a series of sensors are present on drones to support, for example, topographic analysis (LIDAR, acronym for Laser Imaging Detection and Ranging), for analysis of agricultural and water diseases (through multi-spectral and hyper-spectral sensors), for vegetation examination (with sensors of the near infrared), for water and thermography, for advanced cartography (via optical, visual photo and video sensors) for non-invasive and non-destructive material diagnostics thanks to the imaging terahertz, the same technique introduced recently for airport checks by scanning the passenger body to detect explosives or toxic gases.

Sensors are therefore a market that is expanding not only quantitatively but also in the quality of genre and typology: accelerometers, gyros, magnetometers, pressure sensors, temperature, lighting, not to mention chemical and biological ones. We are a *sensor society* – say Andrejevic and Burdon.

Beyond measurements

Even in this case, however, what is relevant to us is not so much the market size of the phenomenon or the continuous emergence of new technologies, devices and practices, but rather the deep philosophical impact that this sensing of reality produces on two dimensions: *a*) "sense production", that is, the redefinition of key concepts we use to produce knowledge of the world; and *b*) "world production", namely the construction of our reality that derives from the introduction and adoption of new cognitive and sensory architectures and technologies.

Here too, we must start with the classic philosophical question: "what is a sensor"? Even in this case, the answer does not have a shared definition and the perspectives with which to look at the sensors can be different. Summarizing current technical definitions, a sensor is a device that intercepts, records, translates and converts a stimulus/input – which is to be measured – in an electrical signal that becomes a measured and interpretable output (by an observer or an instrument). A sensor, therefore, is used to intercept and measure a wide spectrum of phenomena and physical-chemical and other quantities present in the world: traces of chemical or biological entities such as proteins, bacteria, chemicals or gases, light intensity, movement, position, sound, and related events.

Today, our smartphones and tablets, our wearable accessories, shoes and bracelets, our home appliances and shelves, our homes, our gardens and hotels, our pets, our cars, motorcycles and bicycles, our roads and our cities, our offices, our laboratories and our factories, our forests and our seas contain sensors in different quantities and qualities. And, ultimately, they are sensors and nodes in sensor networks and platforms.

According to the most common and trivial consideration, sensory devices are, to put it simply, "instruments of measurement" to improve the understanding of our reality. In this perspective, the sensory networks we are introducing into the world (personal, zonal, environmental, social). constitute a true scientific revolution as meaningful as, in past centuries, the introduction of the telescope and the microscope: they allow us to see phenomena, events or processes that we would not know about without this mediation.[7] Sensors must therefore be considered as a new scientific research tool: a *macroscope* capable of reading complex, physical, biological, ecological and social phenomena – from the infinitely small and invisible to the planetary large and unobservable.

As Josh Berson showed in his "instrumented life",[8] Alex Pentland with his "social physics"[9] and Jennifer Gabrys with her "environmental computation",[10] sensitized networks constitute an extraordinary magnifying lens allowing to make visible and observable physical, social and environmental realities that we have so far not been able to access. Thanks to these new insights into the world introduced by the sensing technologies,[11] we are now beginning to look with new eyes, record and know our bodies, our society and our planet in innovative ways. But is it just a matter of measuring, recording and archiving the world? Is that it? I would say no.

If we look at the phenomenon of "sensorialization" from a more speculative and philosophical point of view, we can begin to consider the sensor
as a tool to "build the world." In fact, sensors are not merely and principal-
ly instruments of *measure* and *knowledge* of the world, but also – more and
more so in the future – of *creation* of the world. Once the sensors are introduced into the world, in the shapes and scales we have anticipated (and
that we shall expand upon here), they become integral part of that environment, that process, that object, that body, that society. They become,
in all respects, ecology-generating vectors and amplified experiences that
we could term socio-techno ecologies. They must, therefore, be considered not only as measuring, archiving and data recording technologies,
but also as world-building technologies and experiences gained through
the ability of environments, objects, processes, individuals and groups to
feel the data. This is the key point. Via a short-circuit that is constantly feeding itself, this new dynamic evolves from "listening to the world
data" to "transforming data into world" (to sense again and implement
it). This is the way the new ontology of sensors develops and operates.
In summary: sensing is not just measuring and understanding an existing reality, but it is creating and acting upon a new reality. It is not – as
most commentators say – just a matter of preventive diagnostics (medical,
industrial, societal and so on) but a new, future-oriented post-prosthetic
techno-ecology based on data.

Action in the world is not only sensorial and cognitive but also direct,
because sensing technologies are often accompanied by "actuation" technologies. *Actuators* are devices equipped with control and automation functions designed to convert sensor data into action and intervention commands in the environment. These are mechanical or electronic tools (but
also of other nature) that execute commands and instructions on things
and environments. Sensors and actuators are often coupled devices (in this
case they are also called *transducers*) that can sense and act on the collected
data. To give a simple example, a gas detector (transducer) has the sensory
capacity to measure in the air the presence of gas in a room and an actuative
mechanism by which an air conditioner is switched on to counteract the
anomaly accurately. The sensor-actuator pair, animated by data, software
and smart algorithms, works by creating an ontology of constant contact
with the environment, real-time and continuous interaction of perceived
distributed atmospheric manoeuver in the intelligent environment, with
decision-making in automatic and autonomous mode, and – this is another

key element – in many cases without the need for human intervention. As we will see in the chapter on experience, all these elements radically redesign our concept of presence, sensation, attention and perception.

This process of sensing and actuating involves a wide range of phenomena today. We are, in fact, experiencing a series of progressive transitions towards a pervasive sensory instrumentation with sensors everywhere. Today, however, there are three major domains of this instrumentation: body, movement and environment. They cover fields that can be conceptually distinguishable, though strongly interwoven with each other. On these, a more speculative analysis will help us to clarify what we have anticipated so far.

Socrates wears the fitbit

Genomics and neuroscience are mature quantitative disciplines, but today the dominant human data domain is the one related to body measurement and quantification of human behavior.

Medicine and health, media and entertainment, sport and wellbeing, financial and insurance services, and the food sector are just a few of the industries concerned, in various ways, about knowing the data of citizens and consumers. When Socrates proposed the Delphic motto "Know thyself", as a behavior guiding principle, he would never have imagined that new sensing technologies would give a whole new meaning to his philosophical speculation.

"Self-knowledge through numbers". This is the slogan of the movement of the "quantified self" that, born in San Francisco in 2007, embodies today the new discourses and technologies related to the quantification of the body as "data". The phenomenon is defined in many ways: self-tracking, body-tracking, personal informatics, human analytics, lifelogging, life hacking, instrumented life, calculable body (computable body). In some of its forms – as in the case of the quantified self – it has created a people movement and an expanding distributed socio-cultural community.

Beyond the labels that look at the phenomenon from different angles (quantification, customization, tracking, hacking and so on), in essence, the purpose is clear. It describes the digital practices – supported by information technology and increasingly widespread among the population – of monitoring, collecting, quantifying and displaying data relating to the physiological, postural, positional, and behavioral habits of an individual. All this can be done through wearable or portable devices (the most popular today are smartphones, bracelets and watches), implanted or ingested (less frequently in different cases and niches, along with subcutaneous or intra corporal nano-implants) or environmental. There are also systems that develop wireless domotics to remotely detect home residents' emotions by measuring heart rate and respiratory signals.[12] More often, we talk about self-observation techniques associated with passive self-archiving practices ("self" understood as personal, but also as automated and automatic).

Of course, this physiological and neurophysiological computation is possible because the human body is chemical, electrical, mechanical, thermal as well as being positioned in a given place and moving with gesture and displacement. In particular, biosensing technologies combine a biological element (such as saliva, sweat) with an optical or electrochemical physical-chemical detector. The detector interacts with the analyzed substance and converts some of its aspects into an electrical signal that is displayed or expressed in such a way that it is interpretable and communicable.

The personal goals of this sensitivity are diverse and range from those who simply record and track the various aspects of their own person and life to those who, through these self-knowledge practices, want to improve dimensions such as personal health and physical tonicity, individual, psychological and emotional wellbeing, their sociality and relational abilities, productivity and professional performance (with a growing trend from tracking to coaching).

A first important philosophical observation to make here is that "sensing the self" becomes "somatising the self": old practices of introspection, self-knowledge, subject-based meditation are replaced by body-centered practices, continuous quantification, and measurement technologies, according to Abend and Fuchs.

This self-sensitivity has distinctive bio-technological and socio-technological characteristics: *a*) they are digitized and automated practices (self-taught or hetero-detected) that allow unprecedented granularity, pervasiveness, non-invasiveness and continuous monitoring; *b*) personal data is collected, transmitted, archived and made available in clouds (personal portability, but also to third parties for institutional or commercial purposes); *c*) the socialized dimension (from body to sociality: sharing, comparing, influencing, humoring) is increasingly promoted, encouraged and empowered by applications and tracking platforms in relation to the dynamics and the mechanics with which they are designed; *d*) self-knowledge is the prerequisite for the self-improvement practices of your wellbeing condition.

Another important philosophical aspect is that, in this way, "the 'truth' is always already communicated from body to machine, before the subject has said a word".[13] Even in this new, unexpected version of the Turing test,[14] machines reverse roles. We are not asking them to understand how intelligent they are. It is rather they who come back to question us about who we are and we cannot lie (in this case, our body talks for us). Our body will confess to the machine the truth about its state of health and psycho-physical wellbeing. Somatic instrumentation is our contemporary truth machine.

In this perspective then, the practice of the quantified self is both confession and surveillance. The sensor is a personal device and a panoptic apparatus, a confessional and a prison (to put it as would Michel Foucault, often cited in contemporary debates about privacy, monitoring and control). With a machinic novelty in our time. When code, sensors and machines point directly to the body – as we will see in The Datum chapter – the conscious subject is left in a peripheral location. It is no longer just a question of privacy (protection), but of destiny (protention): it is not just who knows what, but who and why anticipate us subperceptually listening to our bodies.

And this is the third key point. Sensing technologies are able to access and build an artificial sensitivity totally or partially inaccessible to human consciousness and awareness (such as my skin's galvanic response, my heart rate, or counting the exact steps in a day, a month, a year). With body sensors, all our bodies become inputs to computation, processing data and acting the world. At the beginning of computation, we typed on a keyboard, then we touched the screens, now our whole body, its gestures and, more intimately, its individual internal organs and its cell biochemistry contribute to computing and nourishing the code. There is a progressive physical vicinity of code to body (centralized, personal, mobile, wearable, tattooed, implanted). To the point that there is already talk about data intimacy and new machine capabilities to know us better than we know ourselves. However, we must also ask ourselves: what happens to our idea of body experience when we penetrate it, or when we wear it, or carry or instill in our body sensors that begin to generate historical data series about our physiology and our behavior?

What is most relevant to us here is that these self-measuring practices –
constantly oscillating between information, care, control, and surveillance
– are changing our sense of body and individuality at a glance and pro-
foundly. It is not – as is commonly thought – just about tools to calculate
the number of steps we have taken yesterday or the sum of calories we have
burned in a stroke or, yet, to measure heart rate levels or to map the urban
path followed to arrive at a particular place. But rather, they are tools with
which we are building our new (idea of) human subjectivity. Let us then
ask what kind of subjectivity is produced by this new phase of "human
quantification", after the Quetelet's average person[15] and after the digital
age of one-to-one marketing.[16]

Personal sensor data is a new evolutionary force that is redesigning the
interface between the body and the world – says Berson – along with a
profound revision of the conceptual matrix of Body and World, in such a
way that it has become increasingly difficult to grasp it with the traditional
tools of historical and anthropological contextualization.

The philosophical redefinition of the body concept becomes even more
necessary when we step up from a practice of non-invasive hacking (*soft
hacking*) to hacking a surgical operation that senses bodies through the
prosthesis implant (*hard hacking*) – as the Forrest biohacker did. With remarkable ethical and political implications: while it was once considered
normal to die because of being infected with the measles virus, today the
introduction of an augmentation operation, which is vaccination, is considered a necessary and fair practice. But where should one draw the line?
Are there problems with the massive introduction of prosthetic objects
that are not immediately needed for medical or repair emergencies? Kara
Plato asks what will happen if ten arms are considered economically more
productive than two.

It is not just a question of privacy in managing big personal data – as
tends to be argued. It is also a question of how we intend to build new
equality and freedom even as human beings are being augmented by sensors, data, codes, and algorithms.

Of course, altering human beings with technology is not a totally new
idea or practice (but "alteration" implies, wrongly, that there is a distinction between human and technology). From the use of psychoactive drugs
to hallucinatory storytelling, different technologies have been put in place
in the past and still today to introduce "other" realities. And yet it was, and
is, in all these cases, temporary alterations, limited in time, to allow us to

access other worlds and without the will to ultimately supplant everyday reality. With their ability, however, to become part of the ecology and environment in which we are immersed, to become pervasive, continuous and capillary, and to interact directly with the sensory system (bypassing consciousness) the new sensing technologies open us to a new era, allowing us to control and alter perception in a much deeper and lasting way. For some, this is a new and positive evolutionary development, but others see the enormous risks and the vulnerability of relying (often unknowingly and invisibly) on extraneous add-ons (platforms, machines, devices, algorithms, data and sensors) redesigning our sensory capacity, our experiential life and our cognitive equipment. Moreover, such redesign and revelation have imperceptibly already begun.

Movements and logistics

We are, ultimately, creatures and creations of movement, of moving ourselves and moving things in the world. Undoubtedly the logistics of handling people, but also objects, goods, content and resources is the second key domain on which the sensing instrumentation is being applied today. Motor sensing and quantification (from human body to environmental logic) requires, however, a deep analysis of movement and, ultimately, also requires really rethinking movement itself, or what "movement" is.

What happens when we translate the physicality of a body movement or object or process into a flow of continuous numbers, data and algorithms? What can we think about physical movement when it becomes a process of data?

"Infrastructure makes worlds, software coordinates labor and logistics governs movement", says Rossiter.[17] It is necessary to dwell philosophically on the centrality of logistics, which is experiencing a season of profound technological transformations. Indications of the arrival of a new "logistical capitalism" are appearing, that could become the central axis of economic valuation.[18] Evolving from the idea of supply chain capitalism to logistical media theory, engineers, urbanists, economists, geographers and sociologists are now impelled to redefine the ontology of the world in the light of a logistics that is ever more instrumental, automated and incomprehensible to humans.[19]

We face a sort of *logistification* of the world, which has become increasingly complex. Consider just three data.[20] In the past twenty years, to do our shopping, we have evolved from traveling about 50 miles a year on average to more than 6,000 (including electronic commerce); the delivery methods offered by retailers have gone from five to ten in the last five years; and 63 percent of logistics managers claim to have a shortage of efficient supply chains. The new logistics is made of voice recognition technologies for the management and preparation of orders and for tracing them via GPS, RFID, radar, biometric monitoring, robots, drones and all this with blockchain in perspective. All these tools today depend on logistical mediation to calibrate work and life in mobility, govern the circulation of objects, and dynamic use of environments. With code, sensors and algorithms, logistics mediate of all these components to make the world operational through movement. Furthermore, regarding the existence of the logistic software that governs this "door-to-door" world, we usually become aware of it only when, for some reason, it stops working or works in an inefficient way. Data and software here have the primary function of extracting value by optimizing the efficiency of the chain or the supply line, Rossiter suggests. When we talk about new logistics here, however, we must make an effort of philosophical abstraction and understand the phenomenon in broader terms than what is traditionally meant.

Logistic mediation operations are in fact events that might seem very distant from one another: the recent arrival of MOOCs (*massive open online courses*); the new autonomous vehicles; the sensorized and robotized assembly chain (*industry 4.0*); the redesign of the energy grid (nano and micro-grids) with homes that become producers, hubs and, in the end, actors in the future energy marketplace; the distribution of information and content through *augmented* and *virtual reality*; 3D printing with its distributed manufacturing paradigm (*3D printing*); the platforms of the on-demand economy of rooms, cars and much more (*sharing* and *on-demand economy*); decentralized protocols of trusts, smart contracts, and autonomous organizations with the new *blockchain*.

From this point of view, classical disciplines such as business and management, military history and economic geography are no longer sufficient to understand more and more programmable logistics, managed through software code and algorithms. For example, we may deem the organization of Amazon's warehouse "chaotic", but its logistics infrastructure has made it one of its major strategic and competitive assets to the point of becoming the world's leading logistic *platform as a service* with its Amazon Logistics Services. Sensors, software and algorithms allow it today to manage the warehouse by randomly allocating objects on the shelves; an algorithm then indicates the fastest route within the warehouse to select in the right order the various products requested by a buyer. The presence of sensors networks and software programs has allowed Amazon to disregard the intuitive and conventional logic of physical storage in homogeneous categories, imagining a *chaotic storage* architecture that has a new, "non-human" logic and intelligence.

A new warehouse logistics mediation with an algorithmic inventory has thus taken the place of traditional "notational" logistics. The latter was deductive, hierarchical, followed by standardized, comprehensible and predictable routines for a human intelligence. The new algorithm, created by using small automated robotic drive units (RDUs), does not require the objects to occupy space in a fixed manner, but constantly redistributes them dynamically, for example, by varying orders, their typology and frequency – according to an order of allocation that is perfectly intelligible for the algorithms that govern it and totally incomprehensible to humans.

This is a clear example of what we said in more abstract way in the beginning, namely that sensing the world not only allows to "feel" the world through the data, and in the example above, to know exactly where the objects are placed in the warehouse space, but to become a part of the new world, of its new ontology: it determines as chaotic the conditions for the world that humans live in (the store in this case) because it is created with counterintuitive artificial logic and intelligence that is perfectly readable by a machine. Following Rossiter, the new sensorialized logistical mediation defines the world through an emergent elasticity of space-time: logistical temporality and spatiality modulate and redesign our mobilized (personal and professional) lives. And, paradoxically, even immobility.

To take another example, think of the case of self-driving intelligent cars. Inserting sensors and actuators with more and more artificial intelligence immediately reconfigures the entire urban mobility – and stationary – system as well as the sense of a logistics city and its time and space. If we think, current (non-smart) cars are now a significant source of space-time waste, the effects of which are in the eyes of everyone: traffic problems, pollution, extra fuel consumption for parking, urban visibility for pedestrians, not to mention human mortality. Say we are now at the parking:[21] on average, a car spends 95 percent of the time in a state of immobility while remaining parked when we are at home, when we are at work, at the gym or at the supermarket. Cars are therefore "hungry" for parking lots and most of them require multiple parking spaces, remaining on average 23 hours on 24. To enable this, cities need to build resting areas, deserted in some hours, and overloaded in others. In some cities (such as Los Angeles and Melbourne) these areas occupy more than 70 percent of urban space. Analysts argue,[22] that if cars became all sensed and autonomous in driving, the need for such huge parking lots would be drastically reduced.

Thanks to the new sensorial, logistical mediation, car data collection is not only an instrument for the car itself, it can also create a digital model of the world, which is technically defined as an *occupancy grid*; suffice it to type "Google car vision" on any search engine to retrieve images of what any of these cars can see: they are the tools that enable the car to create the ontology of the world where it must operate. And in this new logistics the data also come from many other sources that can be cross-linked:[23] opportunistic sensing (for example, data on the use of credit cards in the city, but analyzed for purposes other than the original ones), sensors installed at hoc (for example, road temperature data to assess winter icing and salting dosage setting) and participatory sensing (e.g. social network traffic data used by automakers to exchange information about road conditions).

Environmental sensing

In recent years, the empire of sensors and senses has progressively been expanded to cover the whole world: bodies and objects as much as logistics and movement, even spaces and environments: our homes, shelters and urban landscapes including the ecological niches inhabited by wildlife species all the way up to the instrumentation of the whole planet and beyond, to the extraterrestrial spaces. From the micro to the macro scale, we are building a new sensorial apparatus. The sensors and the data that we are incorporating into the world redefine our concept of environment while at the same time are part of both the new domestic ecology and the undomesticated environment.

We can take the above cases as examples of how we are henceforth living in a world that has become "programmable". Builders, architects and urbanists are keen to use sensing devices in the domestic space to build smarter homes. Interest is also strong in companies and software platforms such as Alphabet (with Google Home), Apple (with Apple Home-

Kit) and Amazon (with Amazon Echo). Some 60 percent of smart home studies now come from computing, mathematics and engineering, 20 percent from medicine and health care disciplines, the remaining 20 percent from researches based on social sciences such as economics, psychology or energy usage and consuming. Consumers and citizens in such studies are rarely mentioned and that, only incidentally; among the researches that cite users most are related to the so-called user experience (which we will return to critically in the chapter on experience), that is, to the needs or the degree of confidence and acceptance of the new home automation solutions.

Little attention is paid to the impacts and ways in which human, digital and material combine to form new manners of being in the world. Generally, these user-centered studies tend to reiterate the increasingly obsolete linear demarcation between human and non-human as well as between material, digital and human; they tend not to consider, instead, the domestic ecosystem as an assembly of sensors, algorithms and emerging interactions. The "augmented" home is a perfect example of the "code/space" concept in which software, data and spatiality become mutually constitutive, dynamically redefining the nature of that space (we will address this issue in more detail in The Datum chapter). For example, if we incorporate a domestic code and sensors to monitor the health of those who live there, when those sensors and actuators are active the home becomes a personal hospital unit for diagnosis and cure. A sensory mattress could detect the symptoms of sleep disorders, while a sanitary sensor could monitor, with a daily analysis, medical values in the urine. Examples could be multiplied at length. Chronic patients, elderly or sick people with particular diseases may benefit from continuous and non-intrusive data and analysis of this nature. Not to mention the reduction of medical costs.

Of course, in a "quantified house", not only do humans need to develop their ability to interact with intelligent home appliances, but those, in turn, need to acquire a sort of numerical imagination, that is, a sufficient level of intelligence and autonomy to be able to decide and act in accordance with the stimuli and information received. For example, they must be able to understand the energy rates in dynamic pricing and make decisions in accordance with environmental conditions such as reducing electricity consumption and reducing costs.

It is no coincidence that the concept of *prices to devices* is now being developed to promote the future ability of home objects to estimate – autonomously, in real time, and in distributed mode – and negotiate pricing with electricity, gas or other utility companies. Hence, the need to start thinking about marketing aimed at smart objects (not humans): with a neologism, I call *markething* this new way of marketing. Once the objects in the internet of things are enabled to evaluate, negotiate and decide autonomously (a food reorder fridge, a car that evaluates insurance, an industrial machinery that requires control), companies need to set up strategies and marketing practices that are no longer centered solely on human beings. Domestic objects must be able to construct quantified models of human behavior to be able to act according to habits and preferences. In all this, as you can guess, the boundaries between who or what is being controlled in a given situation are beginning to blur – as we have observed previously.

A quantified home is also an "automated" home,[24] based on the idea that users are not interested in monitoring directly the control of, for example, energy consumption levels, but rather having the objects themselves assessing and adjusting the levels of fruition according to the desired or necessary domestic requirements (with a so-called *set and forget* strategy supporting a remote alert and override capability).[25] Is it perhaps a somewhat deterministic view of the intelligent house that it excludes improvisation and opportunism (in cases such as when, in general, a washing machine is put to use only when one can and not specifically when the energy costs less) and with a typical utilitarian connotation (goal-oriented), or enhancement, which excludes pleasure and fun.

The philosophical and cultural implications of this technological sensing of domestic space will be profound, though today not yet visible. A technology home will probably be a space to renegotiate a gender characterization over a traditionally still feminine home; it will be a context in which the separation between work and leisure will be even less pronounced, an environment in which the mediation of the code will direct those living spaces to the future rather than to the recorded past, a platform that will dialogue as a smart node in new productive energy networks. That these transformations strengthen and resolve gender or social inequalities is an issue that will, sooner or later, have to be dealt with.

Domestic space, however, is not the only kind to be sensorialized. Sensors are progressively being used also in ecological disciplines. There are many planet-wide projects distributed between terrestrial, marine and aircraft environments that use environmental sensing technologies. Sensors are used to monitor climate change, human settlement processes, the behavior of animal species as well as the relationship between all these aspects. In the end, then, we are witnessing a sensory revolution of nature itself.

What changes can occur to the thinking and philosophy of nature when, for example, sensors and actuators are used to trace animal species or grafted and embedded in savannah, forests, and oceans? Some argue that even while living on the same planet, in reality we experiment and live in contingent worlds and that it is our anthropocentrism that makes us believe that all beings sharing the same meadow (a cow that ruminates, a duck that flaps its wings, a butterfly lying on a flower or a microspore floating in the air while we are lying down, enjoying the sun) perceive and live the "same" experience. Each species (in fact) is closed in its own world niche by having access to a species-specific range of stimuli, signals and information that only indirectly and unconsciously reconnect to an inter-acting ecosystem. With code and data, we can now access these multiple worlds, at the same time creating new techno-ecologies.

Along with the better-known industrial internet, we are also entering the era of the *animal internet,*[26] an era that has the potential to open a new chapter in the long evolutionary history of the relationship between human and animal, domestic and wild, right from the capacity, built up with sensors, networks and data, to give new meaning and new awareness to our idea of nature as well as a new orientation to ecological thinking. Pschera observes that, in fact, in recent decades, we have developed a closer proximity to animals that are being depleted, merely compensating for the loss of our direct and sensory relationship with species. Consider for a moment bird watching (solitary practices centred on a merely aesthetic experience), zoo and safari parks (theatralized and anesthetized reconstructions of our ancestral encounters), domestic pets (decontextualized, desexualized or otherwise modified animals destined to become creatures for the family, for home furnishings or used as symbolic fetishes for the service of human socialities and rituals). Of course, we are not imagining here, as many others, a kind of naïve return to some original nature, to a state of nature that actually never existed.

The first philosophical question related to the environmental sensors we are setting for ourselves is not then whether we can return to nature, but if we "can see animal nature differently". And here is the *animal internet* that is about to change nature, as the human internet has changed society. Animals are themselves extraordinary sensors: if we can hook up these natural sensors with a network of digital sensors, the world's intelligence capabilities are to be multiplied and amplified on a meaningful (and world-wide) scale. Animals become sensitized nodes and platforms through which some of the most complex ecological issues can be addressed and remedied. Such technical approaches are already being introduced in precision farming, which employs the big data produced by sensors implanted in cattle herds to monitor, for example, the state of fertility, and calibrate the fertilization intervention most likely to be successful (this, of course, opens up the question of animal rights and that of the ethics that should be applied to their instrumentalization for economic endowment).

What we need at this point is a new ecological philosophical look that allows us to see this world-wide programmability on a planetary scale. But, as we have learned, it is not just a matter of seeing nature in a new way: it is a new ecological thinking that has to emerge from the application of sensors to the whole planet, animals and the environment together, and that is recognized in the "becoming environmental of computation",[27] as, Jennifer Gabrys, the philosopher of sensorized ecologies, says.

In this perspective, the environment becomes an ecosystem that is instantiated and evolves through sensing and mining technologies. The environment is not only something that we measure through sensors. Sensors and environment become one. The "programmability" we incorporate in the environments through the presence of sensors, code, and machine intelligence becomes fully part of that new ecology. Sensing technologies are constituent of meaning and of the world. They too "experience" through new non-human perceptions and through data. Gabrys writes:

> Environmental monitoring through sensors networks is the practice of making – and not just capturing – environments as processes. Sensor networks are tuned to distributions of relations [...] Environmental monitoring through sensory network mobilizes and concretizes environments in distinct ways by localizing computational processes of sensing within environments and across more-than-human experiences, while also articulating those relationships through algorithmic processes for parsing data. As these processes inevitably compose the possibility of sensing environments in particular ways, they also in-form which participants and participatory modes of sensing register in the perceptive processes of sensor technologies.[28]

Sensing technologies applied to the environment further displace the speculative horizon of what is meant by experiencing the world. In fact – as we will see in The Datum chapter – this new environmental perspective introduces a changing concept of experience beyond the traditional philosophies of sensitivity. The concept of experience is changed not only by the fact that new sensory registers are accessed (i.e., information spectrum so far excluded from human knowledge), but also because of the multiplication of possible subjects (not just humans) that can be drawn to these new levels, scales and data of widespread experience, and the sensory relationships between them and the environment itself.

For example, let's take a very common environmental sensor today: the cameras we monitor, safeguarding against criminals, spotting incendiary actions, or checking seals to stop potential oil leaks from underwater transport networks. By operating them more and more as *visual sensors*, the images of these environments, active in streaming and recorded, become part of the environmental system and work as technologies that generate – through continuous, objective, distributed and non-human sensory data – new techno-geographies of the experience. Human intervention and participation in these new sensing technology contexts are not necessary: there is a production of sensorial experience regardless of the presence of direct human activity. Intelligence and data analysis systems are activated automatically and artificially to produce meaning and action from the collected data.

This new sensory mediation invests and redesigns the concept of experience by making it evolve – as we will see later – from the human dimension to, precisely, a comprehensive environmental scope. Digital media theorists speak of "atmospheric media" in this regard to signal a transition from *media-object* (media focusing on capturing human attention through objects) to *media-environment* (techno-ecologies designed to function autonomously, interacting, pervasively, with human and non-human agents).

We saw in the first chapter that, through the interrupt function, code and computation have been opened to the world. In this second chapter, we have analyzed how, through sensors, the code has become coextensive with the world, an integral part of it. In the next chapter we will see how, through algorithms, the code begins to "think" and create a smarter world with new forms of intelligences that – provocatively – we have called alien.

Notes

[1] In neuroscience, "writing" and "reading" indicate how our senses collect inputs from the outside world and pass them to the brain, and how they interpret the instructions to return from this investigation.

[2] The story of the biohacker Forrest is described in detail in Platoni (2015), who reviews what is happening in frontier labs with respect to the transformation of our five senses by technology. It is a journey into the present and future prospects of bio-hacking (soft and hard).

[3] BI Intelligence (2016).

[4] Lipson and Kurman (2016).

[5] The data is resumed by a 2016 research by Chetan Sharma Consulting and by a Statista survey (2018).

[6] Chamayao (2016).

[7] Of course, software and sensors mediate the relationship with the world in a given manner. Let's take the case of the microbial display in an electronic microscope: the first mediation intercepts and translates the sensory stimuli underneath the human senses in machine language (from inside out) and the second interprets the data translated into machine language and makes them useful to the researcher on a screen or other output tool (from inside to outside). Errors may occur in one case or the other however, and/or in any case the software and its algorithms may and must interpret the information by executing (for their choice and/or data constraints) the operative selections. See also Berry (2011).

[8] Berson (2016).

[9] Pentland (2014).

[10] Gabrys (2016).

[11] This is a topic that is totally underestimated: that of the relation between the residual trace of an event and the reconstruction of the event that generated it. All those who deal with big data are usually not professional historiographers or epistemologists, and are not familiar with the methodology of reconstructive work on the track that is precisely the locus of the historiography. Moving from the trace to the event that generated it is not a trivial action either immediate or devoid of epistemological critiques. In thermodynamics, the trace is the residue of an event. In an operationistic, historiographic perspective, tracing from a historical (human or non-human) residue to the human or natural event that generated it is a scientific operation involving a series of historical reconstruction operations. This historiographic sensibility is usually absent from the common approaches of the big data that take for granted (*data as given*, in fact) the starting point of the analysis (for example, the digital traces of a human behaviour or social habits) to concentrate exclusively on its analysis and interpretation. For an operational view of historical reconstruction, you can see the essays of historiographer and methodologist Luigi Zanzi.

[12] Project of CSAIL.

[13] Hong (2016).

[14] The Turing test is used to test the artificial intelligence of a machine. We will analyze it in detail in The Algorithm chapter.

[15] Belgian statistician and anthropologist (Ghent 1796 – Brussels 1874) Quetelet is remembered for having carried out, among the first, quantitative studies of human and social phenomena, discovering the existence of numerous regularities expressed with mathematical formulas. He formulated the well-known theory of the "average"

man, according to which the physical, intellectual and moral type of a population is identified by an abstract individual in which such characters assume a value equal to the arithmetic mean of the values of the various characteristics possessed by all individuals of the population considered.

[16] One-to-one marketing is the strategy of the network economy that uses internet communication tools, together with sophisticated data analysis software, to identify customers individually, differentiate them, interact with them so as to customize the offer in an extremely specific way.

[17] This is how Rossiter begins his essay (2016) on the new logistic theory. Rossiter discusses the emergence of logistical mediation in its material and design components enabled by operating software systems within global logistics industries. The analysis critically deconstructs the new logistics worlds and the logistic knowledge that impact on the dimensions of labor and economic productivity.

[18] Harney (2014).

[19] Lecavalier (2016).

[20] The data are taken from a research by Conlumino Research in 2016.

[21] Lipson and Kurman (2016).

[22] Naturally, the disappearance of car parks means the loss of monetary revenues deriving from the payment of parking services. With this in mind, the introduction of new technologies requires a systemic approach and some consideration for the complexity of the social and economic world that will have to redesign and rethink.

[23] Ratti and Claudel (2016).

[24] Strengers (2016).

[25] *Set and forget* indicates the practice of configuring the smart device at the beginning of its activation so as to avoid having to return and monitor it over time.

[26] Pschera (2016).

[27] Gabrys (2016), p. 8.

[28] Gabrys (2016), p. 23.

The Algorithm

Exploring/Exploiting Alien Intelligences

We, as humans, still have cognitive limits. Increasingly, as we build technological systems that are ever more complicated and interconnected, we become
less able to understand them, no matter how smart we are or how prodigious
our memory, because these systems are constructed differently from the way
we think
S. Arbesman, *Overcomplicated*

Kasparov returns to move

In May 1997, the greatest chess player in the world became a symbol in
the history of technology: the icon of the defeat of the human mind at the
skills of an intelligent machine. In 2017, on the twentieth anniversary of
the first historic checkmate defeat by Deep Blue, IBM's supercomputer,
Garry Kasparov has once again recounted that experience in his new book,
Deep Thinking, together with his idea of the human future of artificial intelligence. In these twenty years, continuing to play both with and against
machines, Kasparov says he has learned a lot about the vital relationship
that can be established with these our artificial creations. Reading the
book, many of the stories told enter the lively current discussion and debate around artificial intelligence and its return to the world scene as a hot
topic after a long period of frozen interest.

What is known as the "winter season" of artificial intelligence may now
be behind us. A cyclically returning period of disaffection seems to be
closing. It was partly provoked by catastrophic failures, partly by lack or
insufficiency of funds or of interest for the research in the scientific and

industrial development of artificial intelligence. And indeed we are witnessing today a growth of popularity that, for many, is reaching the peak momentum and media hype. To give just one example of this excitement phase, recently and for the first time, Gartner introduced machine learning in its projections on emerging technologies, as one of most popular fields of research and application of artificial intelligence today.

Pessimism and boredom (not to mention fierce internal wars between schools of thought and practice in research on artificial intelligence) have not, however, during those years, prevented universities and companies from continuing – without much enlightenment – to work on the construction of an artificial intelligence. Enthusiasm is the cornerstone in all industries and involves innovative startups and companies with a long history, with academic researchers and developers of solutions for the market.

"At the rate AI technology is improving, a kid born today will rarely need to see a doctor to get a diagnosis by the time they are an adult," says Alan Greene, scientific director of Scanadu, a startup that designs medical devices strengthened by artificial intelligence.[1] The statements of Sundar Pichai, CEO of Google, are similar. Pichai recently stated his intention to transform the company from *mobile first* to *AI first*, a company that thinks no longer in terms of design of services and platforms or exclusively of strategies focused on mobile technologies, but on applications and interfaces guided by artificial intelligence. In its strategy, the artificial intelligence is connoted as the core transformative asset of the company. According to the futurologist Kevin Kelly, in 2026, the main service of Google will be artificial intelligence and not just searching activities. In view of this, it is therefore not surprising that the major digital companies (those that have reached top market capitalization in 2018) are today very committed to the development of specific areas of artificial intelligence. From Facebook to Google, from Apple to Amazon, from Microsoft to Alibaba and Tencent, to name a few, all are investing a lot of human and financial resources in artificial intelligence in its various forms (but mostly in deep learning) and with different levels of objectives, acquisition strategies and/or infrastructure development.

Google, for example, through Google Brain, but above all DeepMind and other research groups, even with investments in open software infrastructures like TensorFlow, has supported software projects involving AI and deep learning from a few dozens to over 4,500 (Q4/2017). Or again, through FAIR (Facebook Artificial Intelligence Research), the research group it founded five years ago now boasting over sixty engineers between New York, Menlo Park, Paris and Seattle, Facebook intends to position itself as a leader in long-term research on artificial intelligence. And recently, Facebook, Amazon, Alphabet, Apple, IBM and Microsoft have created an initiative and a strategic alliance (Partnership for AI) aimed at promoting research and the spread of artificial intelligence to the benefit of the economy and society as a whole. In the next five years,[2] investments in artificial intelligence will grow by an average of 50 percent year after year to reach over six billion dollars in 2020. What are the objective reasons for this new spring (after several winters and other short-lived springs)? Major computational capabilities for programs and software, sensors and data in increasing quantity and sources and performing algorithms are the three vectors that, combined in unexpected ways, are putting the ancient dream of artificial intelligence at the center of current technological discussion and innovation. And, therefore, after the long sixty years (1956–2016) spent since the workshop that John McCarthy organized to present and formalize it to the world, the field of artificial intelligence seems to have finally found its moment of revival and renewed attention.[3]

To date, the ways in which analysts and business leaders imagine this pervasiveness of artificial intelligence in our lives are quite different. Would it be a kind of HAL 9000 mega computer such as presented in the film *2001: A Space Odyssey*, or the birth of the super mind of the Singularity?[4] More likely, it will be a cloud service as a commodity made available to businesses and individuals, at low prices, reliable and distributed like electricity, gas or water. We do not know exactly the forms that it will take, but the prospected future is that of a society artificially made smarter in all its dimensions. As Kelly says, take just about anything, add artificial intelligence to it and the job is done.[5] But then again – as someone more worried would say – take a job, add artificial intelligence to it and you will have lost a job. Doubt remains strong whether it is about assistive technologies (to support human beings in difficulty such as children, the elderly or the sick), augmentative technologies (to increase human capabilities beyond their maximum expression) or substitutive technologies (with the aim of eliminating humans from the performance of activities). Enthusiasm and concerns are divided about this new spring of thinking machines. But is it really like that? In what way will they create a smarter society?

Inhuman, too inhuman

For some, Deep Blue is just a super calculator and has nothing really smart: just brute force can process 200 million positions per second to generate all the potential solutions for eight next moves in a chess game. And, there are those who, paradoxically, say that artificial intelligence is precisely the thing that computers do not possess. There is a need for philosophical clarification at this point: can machines think?

It is hard to say. Turing himself, the father of computational and artificial thought, aware of the difficulty of answering the question, proposed – in jest – the imitation game, which later became well-known as the "Turing test".[6] So, let's see how that game plays out, which later was taken seriously by everyone.

That it was called imitation is not perchance. There are different variants, but basically a human communicates and converses on various topics with an unknown interlocutor who replies, through a screen, by typing. The hidden interlocutor must simulate the conversation and the answers that a human would give; if, at the end of the conversation, the human will not be able to say whether the unknown partner is a person or a machine, then the machine that answered can be said to be intelligent. Easy, right? But of course, as usual, the devil (the machine?) is in the details. One detail for all: to overcome the test, the hidden machine – says Turing – can lie. And, in point of fact, it is called upon to do so by the implicit rules of the game: suppose it responded to the interlocutor that it really was a machine, it would be discovered immediately and therefore would be considered unintelligent.

If we think about it, the measure of intelligence would be here in the capacity to lie, to simulate and to seem like a human (surely it is one of the human intellectual abilities, but perhaps not exactly the one we would like to develop in a machine). In any case, despite various attempts and some claimed victories later denied, to date no computer would seem to have passed the test. Some researchers take the event as imminent, others consider it impossible. Ironically, even as we are waiting for a certified winner of the Turing test, conversely, today there are platforms and architectures of digital services (machines, therefore) that vice versa asks us every day – for protection and security that they carry out for our benefit – to prove that we are human.

So now, there are machines that make us pass the Turing test. Think of all the times we are called to solve a *captcha* to demonstrate to a system that we are humans and not bots or artificial agents.[7] And in this case, in order to be identified we cannot lie, otherwise we cannot access the services. Or, as for the wearable devices of the quantified self we talked about, the machines interrogate our body daily and tirelessly in search of the anticipated truth about our health, our state of well-being, our behavior and our mood.

Turing's puzzle and its subtleties are very much liked by philosophers (less by those who are actually called to build intelligent machines that do things), but at any rate – should we ask – would the test be sufficient to really define an artificial entity as intelligent? Supposing that a machine was able to match the performance of a human intelligence: would we be willing to recognize real understanding, creativity, conscience (together with morality, ethics, law, freedom, will)? These, obviously, are not scientific, but purely philosophical questions.[8]

Historically, research on artificial intelligence and the construction of thinking machines has oscillated between two great philosophical approaches: *a*) constructing machines that imitate the human capacity for problem solving and *b*) constructing machines that do not necessarily replicate human cognitive mechanisms. McCarthy had been explicit at the time: artificial intelligence comes from computer science and serves to solve problems, not from psychology more interested in understanding how the human mind works. After all, an airplane flies without necessarily replicating the flight of a bird. In any case, human intelligence functions as a comparative parameter because we believe it, wrongly or rightly, to be the maximum expression of this capacity to understand and act in the world in the most appropriate, adaptive and transformative way.

And yet, there are specific activities and tasks that machines perform far better than humans: in this case, therefore, the choice to use human intelligence as a comparison does not help us much. The current definitions of "artificial intelligence" will require us to specify what is "intelligence" and what is "artificial". Unfortunately, neither one term nor the other seems to have in turn a clear and univocally shared description. In the case of the former, it would be necessary to agree on what is meant by intelligence; in the case of the second, if by artificial we mean mechanical, many doubt that what the machines today are able to do has anything remotely to do with what we usually indicate as intelligence. Undoubtedly, the measure of human intelligence is anything but simple and uncontroversial: intelligence indexes (IQ), markers of mental abilities, aptitude tests, still contend for the title of those who are better able to identify the definitive indicators of human intelligence.

The current definition, whose root we can trace to McCarthy, one of the founding fathers of the discipline, identifies the essence of artificial intelligence with the idea of creating a machine capable of behavior that we would call intelligent if we saw it acted by a human being (and here, we are back with Turing). Or, in other words, artificial intelligence tries to build computers that are able to do what the human mind can do. The human mind, however, according to many, has multiple dimensions. Some are normally described as "intelligent" (reasoning, for example), others are psychological or physiological (perceiving, predicting, planning, having motor control) and they are involved in ensuring that a human being can perform efficiently tasks that are deemed intelligent. But then, to further complicate the matter, why address only the mind? In all this does the body not matter?

As we have anticipated, many machines are capable of carrying out tasks that a human being would not be able to bring to term and, nevertheless, we perceive this behavior as having some intelligence, although not human. That a security program can block a computer attack in a server's access to a server's data in a thousandth of a second by detecting abnormal behavior, that a tsunami alert system can send an alarm by intercepting imperceptible signals from the ocean floor, that a cancer research program may propose a new medicinal composition having discovered a new molecular reaction to a therapeutic treatment, as a rule we consider all these examples as expressions of systems and intelligent applications in an "artificial" way.

According to others, intelligence is to make appropriate generalizations within a reasonable time on a limited set of data: the larger the domain application, the faster the conclusions reached with minimal information can be defined more intelligent. Learning, as a general process, is in fact the capacity to elaborate temporal sequential generalizations by learning progressively from experience to prepare future actions – hence new generalizations, new verifications and adjustments, and new projections of future action.

Still others point to the fact that intelligence is also how we make mistakes, how they are interpreted with respect to expectations and how they impact on the progress of knowledge.

In the face of similar difficulties, many have begun to look for less philosophical, more operational and circumscribed perspectives and to look more at the concrete projects of artificial intelligence (whatever it is) that are realized concretely and starting from there. This stance recalls the fundamental distinction that we must always keep in mind when we talk about artificial intelligence: that between a "general artificial intelligence" (strong AI or AGI) and a "narrow artificial intelligence" (weak AI or NAI). The first one aims to emulate human intelligence (to create machines that are actually and consciously intelligent), the second to simulate it (to create machines that act as if they were intelligent by simulating the human mind). The strong version, of course, is the most complex one to imagine and verify.

Let's take the game of chess that is considered, wrongly or rightly, a field of application and testing of artificial intelligence.[9] Technically, the game of chess is: a game with perfect information (everything is visible unlike, for example, the game of cards where, some being hidden, the information is imperfect), zero-sum (one move winning for one player causes the other to lose), deterministic (there is no place for what we call luck). It is a game that, at the beginning of artificial intelligence, a machine was not expected to win competing against a middle-ranking chess player, let alone against a champion.

In 1997, thirty years after those pessimistic predictions, Kasparov lost the decisive game against a machine: for the first time in history, the best chess player was a machine. What happened in these thirty years to allow a machine to win? In short: the sophistication in the design of the algorithms, the increase in power and speed of calculation, the improvement of heuristics for approximate solutions, and the application of the learning tool to adjust parameters from experience. The crossing of these elements has allowed the machines to overcome human playing abilities.

Of course, not all games have the three chess features mentioned above and there are games where humans still stand to win. How are humans winning? It is not a simple answer: we can look inside the algorithms of the machines, but the human mind, although explored in recent years by neuroscience and psychology, in many aspects is still unknown in its operation. In the case of chess and other games,[10] looking at how the champions play has emerged as a typically human ability to: *a*) play/think for abstract patterns of moves (not for a single move) and *b*) play/think for anticipation of patterns of emerging moves (not just for the patterns present).

Machines and humans show, therefore, various differences in understanding and interacting in the world: from how they play chess to how they make a medical diagnosis and to how they see the world.

Let us take the latter theme – vision – to make a second empirical example. Although "seeing" for a machine does not have the same meaning it has for a human, its visual skills have achieved excellent goals: facial recognition with 98 percent success, the interpretation of cursive writing, the detection of suspicious behavior in parking lots. But in many cases, still, the machine must know what to look for and what it is looking for must be in specific conditions: faces not upside down, not profiled, not partially hidden.[11] We say in many cases, because also in this area the technological development is progressing rapidly and, in a recent episode, it would seem that the neural networks of Google's computers have succeeded, after three days spent watching millions of images and connections not labeled and not guided, to learn to distinguish human from animal faces.

Although we humans still do not know how we can do many of the things we do, we have understood some indications of our current limitations and those of extended artificial intelligence. In particular, two: implicit assumption and sense of relevance. Understanding the world comes, above all, from the context or frame information we possess (and that most machines do not yet have today) and also involves our ability to grasp the relevance and pertinence in judging a situation (and even in this, to date, computers are not illuminating). Calculation and judgment – they say – are two different things, as we can understand by listening to the decontextualized and sometimes amusing responses of Siri and Watson (a recent empirical test compared four artificial helpers such as Google Assistant, Apple's Siri, Amazon's Alexa and Cortana from Microsoft, with a set of questions, from health to sport to travel, evaluating the performances).

Humans and machines have differences. But often we take these differences for granted. Humans consider themselves more creative, intuitive, emotional (but, of course, we do not know what creativity is, or intuition, or emotion); vice versa machines are seen as lacking in consciousness and emotion, mechanical, repetitive and soulless (although Deep Mind creatively invented a move never played before in the millenary history of Go game). Some of these differences are objective and measurable, others depend, however, on how we look at machines, especially in order to preserve and characterize our distinctiveness as humans. Hui[12] says that, despite everything, we continue to believe that the human is the only subject capable of imagination. But, to the point, it is just that, a belief.

At the same time machines and humans also have different limits. Physiological, psychological, emotional limits and social conditions for people; for machines, algorithmic limits, computing power, complexity in being in a world made for humans (still now, but less and less so): to give a simple example, we think about websites and the need to design them not only for the humans who use them, but also for the artificial bots that index them and put them in search engine rankings. However, the differences are narrowing and the limits of the current machines are being overcome. On the other hand, their skills increase significantly: to take a recent example, Google's DeepMind, working with artificial neural networks or ANN,[13] has developed selective memory skills by learning to understand what knowledge to keep in its learning path.

In principle, but also by experience – as we have learned in recent years – we should bear in mind that there are no processes and activities that cannot be artificially and algorithmically generated. Even more: we live more and more immersed in the algorithms: an eco-system that grows by building our world ontogenetically. To the point that if today all the algorithms stop working it would be the end of the world as we know it, says Domingos.[14]

Artificial intelligence therefore pulsates with an algorithmic heart.

Ontologies as algorhythmies

Not surprisingly, in a provocative manner in the introduction to his essay *Digital Stockholm Syndrome in the Post-Ontological Age*, Mark Jarzombek, professor of history and theory of architecture at MIT, asked us to rethink the contemporary world as a result of the meeting between the ontological and the algorithmic, between what we believed to be the world and the new reality that the algorithms are creating. For those who might have taken a course in computer science some time ago, talking about algorithms meant dealing with the design of more efficient software programs, manuals and complex articles full of abstract forms, mathematical problems, all seasoned with compilers, libraries, specifications and documentation, operating status of machines and devices. The algorithm, however, is certainly a technical object, but also a form of discourse, professional practice and subject matter, today, of public discussion.

All of a sudden, from discourses and niche practices, algorithms have arrived at the center of public debate, as familiar as collective, positive as negative (and above all negative, these days). In fact, algorithms do not enjoy good press today. Some case studies – we could define them as algorithmic discrimination – have reached the mainstream media and the consequent political discussions, raising the question of fairness, equity and social value of the use of algorithms that filter, recommend, prioritize and classify data and information. To give two examples, in a recent experiment, by typing in a search engine "professional hairstyles for women" and "non-professional women's hairstyles," the algorithm extracts in the first case photographs of white women and in the second, black women although both professionally curated. Another recent case concerns Amazon and the algorithms that determine the prices of products and the presentation of offers to consumers. Bezos claims that Amazon's algorithms are objective and focused on consumer benefit, while some consumers and researchers have raised doubts about the correctness of algorithmic pricing policies.[15]

Algorithms do discriminate, then? The critical thought of Marxist lore speaks, more explicitly, of "algorithms of capital",[16] indicating a politically oriented interpretative way, in which the algorithms are the new exploitation machine to extract value. Not just a mechanism of discrimination, therefore, but also a mechanism for generating surplus value. Says Pasquinelli: "The conceptual operation that I suggest is to apply the notion of machinic to the algorithms of the digital code to recognize the digital code and software programs as a form of machine in the Marxian sense, as a machine used to accumulate and increase the surplus value."[17] Even the liberalist thought shows, from its perspective, some presage for this new *invisible* hand in the markets (of an algorithmic nature and embodied by super-platforms) that puts into question and at risk many of the assumptions on which mechanisms of law and regulators have worked in favour of economic competition, and against discriminatory dynamics, for the benefit of consumers.[18] To date, researchers have classified the actual algorithmic discrimination cases in two main bias categories: representative (depiction of social identities) and allocative (allocation of economic resources).

Let's start with a more technical analysis of algorithms and then reason philosophically on the point. Technically speaking, algorithms are sequences of instructions that, in a precise and univocal way, tell a computer what operations to perform. In computer terms, an algorithm is a formalized abstract description of a computational procedure. Logic plus control – they say. However, code and algorithm are conceptually distinct as Dourish points out:[19] software programs can incorporate or implement algorithms, but they are at the same time something more than algorithms (the programs also contain non-algorithmic elements) and something less (in the sense that the algorithms are not bound to the material reduction implied by a particular implementation).

Certainly, an algorithm can indicate, for example, how to transform a typical representation of data into another or how to reach an appropriate numerical result for a formula or how to transform data (for example, to put them in a certain order); however, the software program will have to do much more than execute this core aspect of the code. And it is precisely this machine procedural aspect, the execution of the sequence of instructions one after the other, that is differentiating: the software will have to read the files from the disk, connect to the servers on the network, verify the error conditions, respond to user interruptions, view visual and audio signals on the screen, drag data between different storage units, record all of these feeds in log files, verify screen size or free disk space and much more. The algorithm can vary by type and characteristics (combinatorial, numerical, probabilistic and so on). To avoid essentialism, however, it is opportune to evaluate it contextually and philosophically in relation to contiguous concepts.

Let us try, then, to analyze it – as Dourish proposes – in combination with other concepts: *a*) algorithm often means a regime of "automation" and, in this perspective, the algorithm is assigned the dimension of control and governance management more generally; *b*) algorithm is often understood as "code" or, better, pseudo-code or code-waiting-to-happen, as an abstract generality that later (temporarily) will be operationalized in a specific programming language; *c*) algorithm as "architecture" because what we call by that name is often difficult to locate and instead is distributed in million lines of code, divided between components, languages and different systems, activated between pervasive interfaces, local and global networks, operating between assembled architectures and machines; *d*) algorithm and "materialization" in that, beyond the coding, it takes the form of a system that makes it operational like a specific machine or a computer, a specific network, a specific hardware configuration.

These components regulate – Dourish continues – the effect of the algorithm: storage speed, network capacity, instruction channeling, memory hierarchies, all have important repercussions on the algorithm's performance (and of course the inverse is also true).

Code and algorithm are united by their specificity. As for software, the issue of the visibility of algorithms becomes more relevant. Pasquale talks about "black box society."[20] Their invisibility would prevent understanding their nature and, in case of discrimination or manipulation, their eventual removal or mitigation. A report[21] by the White House in August 2016 drew attention to the potential risk of algorithmic practices with respect to civil rights and individual or social groups discrimination for racial, economic, sexual, religious or other reasons. The public explosion of Cambridge Analytica critical case in early 2018 (which involved Facebook in a very large media scandal related to data abuse and manipulation risk) points out to both dimensions of data "privacy" protection (to avoid abuse in the present) and people "destiny" protection (to avoid manipulation of the future). The social awareness of the necessity of more transparency and accountability for this algorithmic darkness has progressively grown since 2016, demanding a pervasive and constant fairness and control against opacity and invisibility. Not an easy mission, however.

This opacity is linked to the same dimensions already mentioned regarding code, to which we must add a very significant new one.

The first reason lies in commercial or institutional protection (companies such as Google and Facebook, but also all those that use proprietary algorithms, including states and institutions, do not easily allow access to the algorithms that are at the base of their services or procedures). It is in fact an intellectual property defined as a corporate or an institutional asset and, consequently, held in industrial or governmental security and confidentiality.

The second reason lies in the professional specialization. Understanding complex algorithms requires highly technical knowledge that presupposes education, training and specific skills, an issue that is further exacerbated by the fact that such skills are usually missing or completely inadequate.

The third motivation is extremely relevant and results from a critically important technological innovation: the automated learning experience (machine learning algorithms work by gathering, in an unpredictable and unknowable way, data from how the system is being used). When my bank, through an algorithmic procedure, identifies as suspicious an transaction made with my card and suspends it for the sake of security, it is not always able to explain completely why the algorithm has marked that use as suspicious. Even the legitimate request to enact the transparency of the algorithms, allowing citizens and consumers to understand the logic and the dynamics with which they work to classify, filter, suggest and ultimately understand and interpret the world, although necessary, may therefore not be sufficient nor satisfied. On the other hand, we should also remember here, for comparative completeness, that a certain degree of opacity is present even in the sensory and cognitive modalities with which humans read and interpret the world. Often, we do not know exactly how we, as humans, can recognize an image or reconstruct a meaning or intuit a solution.

In any case, we are moving towards an awareness of the need for accountability and auditing of the algorithms (i.e., a responsible, shared and more transparent knowledge). Some analyses[22] show that algorithms do discriminate, even if and when there is no discriminatory purpose in the intentions of their developers. Algorithmic bias can depend, for example, on the databases used or the attributes chosen as variables for correlations. They can amplify existing discrimination by reinforcing stereotypes and prejudices as much as creating new and unexpected ones. On this point, a recent article in the scientific journal *Nature* supported the urgent need for greater transparency and algorithmic symmetry. And with good reason: in fact, we are entering a world where algorithms are no longer simply instructions that must be executed, but have become performance entities that select, evaluate, transform and produce data and knowledge, in a deterministic or explorative manner.

The ontological and the algorithmic co-emerge, as we have learned to recognize. As with sensors, they are no longer just tools to perform a task, but become a component, at the same time material and abstract, which enables the automated design of our experiences. They are therefore "actualities", defined by the continuous acquisition of data in a process of constant computation of probability.[23]

The algorithm is called upon to scan and retell the daily dynamics of our personal, private, public and professional lives. In this sense, we can review and reimagine the root of the word and give it a new meaning beyond its etymological origin linked to the Arabic mathematician al-Khwarizmi. Algorithm, therefore, as *algo-rhythm*,[24] is becoming a concrete element of the dynamics of our lives – whose political and economic aim remains, as we have said, to be analyzed critically and urgently. Recently, the research group led by professor Alex Pentland began to work in this direction, proposing an orientation to an *open algorithms* (OPAL) approach.

A squared automation

While we wait to untie these nodes, in a world of invisible algorithms (but with visible consequences) and unpredictable (in spite of their deterministic and controlling nature), there is one, in particular, that has encircled us surreptitiously.

"You may not know it, but machine learning is all around you."[25] Pedro Domingos, winner of the 2014 SIGKDD Innovation Award, the highest award for data analysis and intelligence, in the prologue of his *The Master Algorithm*, explains the emerging world owed to the progressive introduction of machine learning in our economy, culture and society. Machine learning is today – and will increasingly be in the coming years – at the center of many of our activities as citizens, consumers, customers, professionals and workers. Its presence is becoming widespread, even though we do not notice it. We see its consequences and effects, but do not recognize its origin, nature or dynamics. And yet, machine learning is at work 24 hours a day, 7 days a week.

Domingos writes:

> Your clock radio goes off at 7:00 a.m. It's playing a song you haven't heard before, but you really like it. Courtesy of Pandora, it's been learning your tastes in music, like your own personal radio jock. Perhaps the song itself was produced with the help of machine learning. You eat breakfast and read the morning paper. It came off the printing press a few hours earlier, the printing process carefully adjusted to avoid streaking using a learning algorithm. The temperature in your house is just right, and your electricity bill noticeably down, since you installed a Nest learning thermostat. As you drive to work, your car continually adjusts fuel injection and exhaust recirculation to get the best gas mileage. You use Inrix, a traffic prediction system, to shorten your rush-hour commute, not to mention lowering your stress level. At work, machine learning helps you combat information overload. You use a data cube to summarize masses of data, look at it from every angle, and drill down on the most important bits. You have a decision to make: Will layout A or B bring more business to your website? A web-learning system tries both out and reports back. You

need to check out a potential supplier's website, but it's in a foreign language. No problem: Google automatically translates it for you [...] You find a flight for an upcoming trip, but hold out on buying the ticket because Bing Travel predicts its price will go down soon. Without realizing it, you accomplish a lot more, hour by hour, than you would without the help of machine learning.[26]

This incursion in everyone's daily life serves to account for how pervasive is the age of algorithms (but the needs could multiply for many of the activities that we practice every day from the time we get up to when we fall asleep). This is the age of the algorithmic economy, the algorithmic business, the algorithmic society, in short, that of algorithmic life and so on. In particular, machine learning algorithms have become a daily presence although, as we have said, invisible in their architectures, dynamics and deep technologies.

But what is machine learning? In short, we can say that machine learning deals with building machines that improve with experience (basically, data). Born within the artificial intelligence family, automatic learning has gained a prominent position to the point of almost eclipsing its roots in recent years. While artificial intelligence tries to teach computers to do the things that humans do best, machine learning aims to teach machines how to learn. It is therefore the systematic study of algorithms, systems and machines that improve their knowledge and performance thanks to experience. In synthesis, it's the machine that learns from the data. And this peculiarity has also been submitted to a more abstract formulation: a given software program is deemed to be learning from an experience E in relation to a certain class of tasks T with a measure of performance P if its performance related to tasks T, measured with P, increases thanks to the experience.[27]

Other definitions stress more the concept of predictive capacity for which machine learning is a set of methods for automatically discovering and extracting patterns within data to take decisions in situations of uncertainty and to predict the future. And this is relevant, for example, from the economic point of view because:

Machine intelligence is, in its essence, a prediction technology, so the economic shift will center around a drop in the cost of prediction. The first effect of machine intelligence will be to lower the cost of goods and services that rely on prediction. This matters because prediction is an input to a host of activities including transportation, agriculture, healthcare, energy manufacturing, and retail. When the cost of any input falls so precipitously, there are two other well-established economic implications. First, we will start using prediction to perform tasks where we previously didn't. Second, the value of other things that complement prediction will rise [...] As the cost of prediction falls, not only will activities that were historically prediction-oriented become cheaper – like inventory management and demand forecasting – but we will also use prediction to tackle other problems for which prediction was not historically an input.[28]

They can say that many of the activities and tasks in the future will be redesigned as a "question of prediction." We'll return to this new predictive dimension later.

In any case, why is machine learning relevant? Because over time, computer scientists have created and recombined algorithms, using their results to further create new algorithms. This necessary work has progressively increased the complexity that – says Domingos – is the monster which algorithm builders face every day. This complexity is no longer manageable through human intelligence alone and many believe that machine learning and its variations are the best candidates. The automatic learner algorithms are in fact algorithms that create other algorithms and, therefore, unlike the classical procedure, the task of creating them is entrusted to the machines. It is difficult to explain how, as humans, we know how to drive a car or recognize a handwritten phrase, but with an increasing number of examples (of data and experience) a learner algorithm will have no problem – experts say with optimism – in understanding just how to do it.

At this point, the conceptual link between data and machine learning should be clear. Data is the raw material on which a learning algorithm can grow and improve. The exponential growth of sensors and the consequent availability of big data (of various types: images, sounds, texts, networks, gestures, positions, movements ...) explain the rising relevance of machine learning.

As a general principle, let us take into account – and this is a key passage – that each of our actions in a digital and artificial world develops on two levels: first, the interaction with the machines allows us to obtain what we are looking for and, secondly, this same interaction teaches the machine to improve the knowledge (model) it has of the world and of us. And we must keep in mind that in the future, more and more of these algorithmic models of us will interact, on our behalf and automatically and continuously, with the algorithmic models of objects, companies, brands, and institutions.

Without data, the learner stops learning and therefore loses its relevance. With increasing amounts of data, the machine learning algorithm can face emerging complexity. A metaphor will help to visualize this relationship between data and algorithms. For Domingos, databases, indexers, and crawlers that contain or collect data are like herbivorous animals grazing in information meadows, while learners are hungry super predators who hunt, feed, and, thanks to the herbivores that ruminate in the endless expanses of data, grow to recognize patterns and regularities (pattern recognition). Naturally, the quality of the results depends a lot on the input examples of the feature extraction that are fed to the algorithm (classifier) because it is thus exercised and can train itself to achieve the adequate recognition.

Exponential complexity, increasing data availability and computing power are therefore giving rise to a new automation revolution. If the industrial revolution automated manual work and the information revolution automated the intellectual one, the revolution of machine learning will come to "automate the automation,"[29] automation squared. More philosophically, automation via artificial intelligence will not simply be a feature that we add to a car or a drone – as vehicle manufacturers continue to believe. It will be an entirely new architecture, an environment or a milieu – philosophers would say – in which objects and events acquire an autonomous alien intelligence. We also note that machine learning could further reinforce its capabilities thanks to the arrival of quantum computing,[30] a computation that uses the computational potential of quantum mechanics, in particular on reinforcement learning,[31] one of the three branches of machine learning.

But, just a moment: what about humans in all this?

Are humans still driving?

Are humans still in the loop – as the software engineers say – or are we waiting for a slow but progressive and inexorable marginalization? Will machines really replace humans, first in labor (a topic that is debated at all levels, from the conferences of the World Economic Forum to the daily bar-room discussions) and then, in a sort of domino effect, in many of the things we do now, for example, driving a car?

Meanwhile, the robotic density (the relationship between robotic and human workers) of the automotive industry in the USA is the highest in all industries (more than a thousand robots for every ten thousand humans). Are not only the humans who consume the car, but also the ones who produce it at risk? Marvin Minsky[32] had no doubts about it and he feared pessimistic scenarios: we should be lucky – he argued – if, in the future, intelligent machines still keep us as pets. So, are we bosses, slaves or partners?

The question is not simple, as you can imagine. Many analyses are trying to model a future of work (and not only) with a strong, algorithmic and robotic component (RPA or robotic process automation).[33] Estimates vary widely: some analysts estimate job losses in OECD countries between 45 and 60 percent, other researchers focusing on specific activities rather than work in general, indicate a more optimistic 9 percent. Other studies, paradoxically, point out that jobs that do not have a computational component are those most at risk. In other words, someone who already works with computers and machines is less likely to be replaced by these machines compared to those whose jobs do not involve computers or computational devices. Some argue that, as with other industrial revolutions, the jobs that will be created from scratch thanks to new technologies will balance those lost. Yet others point out, instead, that we are facing a peculiar industrial revolution – different from the others (it is not just a single sector, but all the sectors are changing) – and that therefore historical comparisons with the past do little to help us. And perhaps we should imagine social policies (such as the introduction of universal citizenship income) that will compensate for an economy with little work for humans or work with low or no wage, as happened in the *heteromation* (think about repetitive microtasks, videogame training, social media interactions) according to Ekbia and Nardi. Probably, in a more radical way, a philosophical rethinking of the concept of work is on the horizon. But it is not – I think – just a question of jobs.

For example, let's take a concrete case:[34] that of intelligent cars that drive themselves. Many people are waiting for a gradual transition to autonomous vehicles with, however, the presence of a human to intervene in case of malfunction or unexpected situations. Now, this option that would seem to be a shared, common sense perspective, in reality hides many complex and counterintuitive elements. Some experts believe that the optimal model is one in which humans and machines have a distributed control of the car, but in which the human remains in command and the software functions in a subordinate way. Assuming, of course, that the responsibilities between human and machine are clear, it is argued that the coupling would function better than with either the human or the machine alone. It is in this way, with human–machine coupling, that many of today's commercial, industrial or military applications work. They are based on the principle that machines are precise, tireless and procedural, while humans are better at making decisions about inaccurate data, making unusual associations and interpreting the context thanks to past knowledge. Together they would do their best, respectively.

However, some recent field experiments would show somewhat different evidence in the case of smart cars. The key point is that once the human being delegates his driving activities to the automatic code – and humans are quick to delegate if they trust that they are safe – returning to a command position, from a situation of relaxed attention (which is the purpose of an autonomous car) in the case of an unexpected event, it is not so immediate or safe. Some examples of airborne disasters testify to this human difficulty – being inattentive and distracted by something else or little prepared for driving because it is no longer exercised or unaware of what is happening around – to effectively return to the command post so as to solve complex situations. The hand over between machine and human has produced and could still produce disastrous errors in the future.

For these reasons, Google has decided to embrace a full automation approach, considering that the first tests carried out by them internally and other experiments conducted by researchers and universities elsewhere have indicated greater risks in opting for semi-automation (that is, with the driver returning to the control of the vehicle) than to delegate routing and guidance entirely to the car. And all the incidents that, to date, have involved a Google Car (with the exception of one) are accidents caused by the fact that the car was driving too well or, conversely, by the fact that the drivers around were human, impatient, not attentive to driving rules or speed limits. Therefore, removing humans from driving would seem to be the most plausible solution to the social problem caused by cars. Deaths by car accident, on average, number about a 1.2 million annually: the equivalent of ten Hiroshima-type atomic bombs being dropped every year on the planet.

According to the World Health Organization, car accidents are the leading cause of death in the world within the 15–29 age group, as well as the second leading cause of death in general. To make a comparison: about 183,000 deaths are due to drug use each year, while murders, suicides and wars combines one responsible for about 1.6 million deaths. In spite of this and unlike other social scourges, no civil battle is being waged against cars – say Lipson and Kurman in their inspiring recent book *Driverless*. Naturally, vehicle manufacturers are pushing for a gradual replacement of the car fleet not only because it is a complex and uncertain route, but above all because this allows them to prolong their control of the car industry.

The same logic of using artificial intelligence with humans present only at the configuration phase (setting) and then as mere observers is at the base of the algorithmic trading (algotrading) operations of finance. Several innovative startups are working to improve automated finance in which the human is out of the loop. But is it enough to remove the human from the equation of our future? Artificial intelligence (AI) or augmented intelligence (HCI)?

It is not a recent question. Back in the 1960s, at Stanford, two conceptual perspectives confronted each other from one side of the campus to the other.[35] To put it simply, McCarthy imagined constructing intelligent machines to replace man (to replace human skills with the development of artificial intelligence), while Engelbart worked to build applications to augment human intelligence (to extend human capabilities working on human – machine interaction). The result is two traditions of thought and two communities, two different sets of values even. To revisit this contrast in our day, we could cite the work of Andy Rubin, the soul of the robotic enterprise of Google, and that of Tom Gruber, one of the key designers of Apple's virtual assistant, Siri.

In fact, many of the artificial intelligence systems built at the beginning were developed to support or replace a single user in performing a generally repetitive or difficult activity and in this direction AI has moved significantly. But, more and more, artificial intelligence goes not only towards intelligent machines, but towards intelligent social machines in which artificial cognitive abilities are increased by human computation. By social machines we mean complex sociotechnical systems that co-emerge from the increased interaction between collective intelligences and artificial intelligences and which imply complex cognitive as well as ethical innovations.

Let us take the example of deep learning to clarify the point concretely. A member of the machine learning family, deep learning is an example of a machine learning technique used extensively in the artificial recognition of images, objects or words. The fundamental aspect is that the extraction of the input characteristics of the learner is not made by humans, but learned automatically by incremental degrees of abstraction and classification that are gradually transformed in a non-linear way. Artificial neural networks deal with operations, with layers (even hidden) of multilevel hierarchical weightings and learning algorithms (for example back propagation or backprop), to allow these abstractions.

Computational speed of training and availability of socially produced databases (videos, sounds, images) to train the learner (training set) are some of the reasons for the success of this approach. Using this technique, deep learning is able to recognize in a photo the presence of humans, animals or objects. But the challenges and complexities of cognitive and ethical nature that await us are still enormous. And opportunities are at least equal to risks and vulnerability.

Let's also suppose that a machine learner distinguishes in an image, a child, a dog and a chair and is able to recognize them as such. Let's also say that it is able to answer questions such as: where would you sit down?, who is more likely to bite whom?, in this photo could someone become a lawyer? And let's also say that it is able, by accessing the collective human knowledge, to understand that a dog is more likely to bite a child than the opposite or to know that a dog has never become a lawyer. But, write Hendler and Mulvehill, authors of *Social Machines*:

> What might an autonomous AI answer if asked 'If you could only save one in an emergency, which would you choose'? If it looked at disaster relief data, it would find that animals and people are typically saved in an emergency, so it could reduce the answer to either dog or person. But what other criterion would the system use to determine which to save? How long would it need to make the decision? What would happen if it made the wrong choice? Who would determine that an answer was wrong and would there be some way to coach the AI to make a different answer? If crowdsourcing was used and more people voted for the dog, would the system make that the answer because it had the highest score?[36]

MIT launched an open platform that investigates this last aspect in relation to self-driving machines. They called it "the moral machine." It acts by proposing to humans a series of optional ethical choices in case of danger or emergency, on which it voluntarily asks to speak. We could say that it is the AI key version of the tower game. Who would you sacrifice if you were in these situations: the person who is in the car or the different pedestrians who bypass each other? And to test the choices, from time to time, children, animals, adults, criminals, pregnant women, the elderly and various combinations of these are put at risk. Who would you save, then?

The answers to these ethical doubts will have to be built collectively and rapidly in the coming months and years. Clearly, we are moving towards an artificial intelligence that is called to interact in ways traditionally reserved for human sociability (that becomes an artificial sociality as machines not only self-learn, but also, for example, teach other machines). In doing this, AI, enriched by sensor data and increasingly sophisticated algorithms, is creating a new society and humanity that are potentially augmented in their ability to experience the world. It is thus laying the foundations for a profound redefinition of our human concept of experience. And this is precisely the theme we will address in the next chapter.

Notes

[1] The expression is by Alan Greene, Chief Medical Officer of Scanadu, quoted in Kelly (2016).

[2] Tractica, *Artificial Intelligence Market Forecasts*, Research Report, August 2016.

[3] John McCarthy is the American computer scientist and mathematician who coined and introduced the term "artificial intelligence" in the mid-1950s. In 1971, he won the Turing Award for his contributions to the field of artificial intelligence.

[4] The Singularity is an idea and a philosophical-scientific perspective that imagines and prefigures a moment in human civilization in which machine intelligence surpasses human intelligence. In futurology, a technological singularity is a point, conjectured in the development of a civilization, in which technological progress accelerates beyond human beings' ability to understand and foresee. The singularity can, more specifically, refer to the advent of an intelligence that is superior to the human (even artificial) intelligence and to the technological advances that, in cascading succession, are presumed to follow from such an event. The concept of technological singularity as it is known today is attributed to the mathematician and novelist Vernor Vinge. In recent years, it has been popularized and promoted by Kurzweil who has linked it to the introduction and adoption of exponential technologies. The debate surrounding the future existence of a singularity remains, to this day, open between supporters and critics. For a considered discussion, see Boden (2016).

[5] Kelly (2016).

[6] For a detailed and updated reconstruction of the state of the art see Warwick and Shah (2016).

[7] *Captcha* is the English acronym denoting in the informatics field a test made of one or more questions and answers to determine if the user is a human and not a computer or bot. Typically used is the one in which the user is asked to write down

the letters or numbers present in a sequence that appears distorted or blurred on the screen. They are completely automatic and do not normally require human intervention for administration or maintenance.

[8] Boden (2016), p. 119.

[9] The analysis of chess and other games (such as Go) from the point of view of artificial intelligence is detailed in Hendler and Mulvehill (2015).

[10] On this topic it is useful to compare Hendler and Mulvehill (2015).

[11] Boden (2016).

[12] Hui (2016).

[13] Artificial Neural Networks (ANN, Artificial Neural Network in English) have become an extremely effective tool in the analysis of situations not "predictable" analytically. Artificial neural networks are information processing systems that attempt to simulate the functioning of biological nervous systems within a computer system that consists of a large number of nerve cells or neurons. An artificial neural network is therefore a mathematical/computer model of calculation based on biological neural networks. This model is made up of a group of interconnections of information conducted by artificial neurons and processed by a connectionist approach to computation. In most cases an artificial neural network is an adaptive system that changes its structure based on external or internal information flowing through the network during the learning phase. An artificial neural network receives external signals on a layer of input nodes (processing units), each of which is connected to numerous internal nodes, organized in several levels. Each node processes the received signals and transmits the result to successive nodes.

[14] Domingos (2015).

[15] The text of Ezrachi and Stucke (2016), is fundamental on the more general issue of price discrimination within an algorithm-driven economy. Examples are the sentence-completion algorithms of the phrases (in Google searches), of priority in the visibility of the news (on Facebook walls), of recommendation of the products (in the suggestions of Amazon).

[16] Pasquinelli (2014), p. 97.

[17] Ibidem.

[18] Ezrachi and Stucke (2016).

[19] Dourish (2016).

[20] Pasquale (2015).

[21] Munoz, Smith and Patil (2016).

[22] See, for example, the indications in Hajian, Bonchi and Castillo (2016) on algorithmic bias, on how to do discriminate discovery and imagine a data mining that is fairness-aware.

[23] Illuminating on this point is Parisi (2013).

[24] Miyazaki (2012).

[25] Domingos (2015), p. 1.

[26] Domingos (2015), p. xii

[27] Mitchell quoted in Hildebrandt (2015).

[28] Agrawal, Gans and Goldfarb (2016). This perspective is detailed in a new book published in April 2018 from Harvard Business Press, *Prediction Machines* by the same authors.

[29] Domingos (2015).

[30] A quantum computer is a device that exploits the laws and dynamics of quantum mechanics in order to exponentially increase information processing capabilities.

Hypothesized by physicist Richard Feynman at the beginning of the 1980s, instead of the conventional bits, the binary information units, conventionally indicated by the digits 0 and 1 and codified by the two open and closed states of a switch, the quantum computer use *qubits*, to reflecting the quantum states of a particle or an atom that can exist in a superposition of other quantum states. Extending the possibilities for information co-ordination enormously, we can face extremely complex problems. Among others, Google and IBM are working hard to create quantum computational machines.

[31] Machine learning paradigms are divided into three groups: unsupervised learning, supervised learning, reinforcement learning. Each of these, starting from a specific experience E (encountered or selected) and given a specific problem T, reaches a certain cognitive performance P (accuracy, relevance). In the case of supervision, examples of behavior or experience are provided to the a priori learner; in the case of non-supervision, there is no trainer that provides correct answers to the inputs; in the case of reinforcement, the learner interacts with the environment by receiving a positive or negative reward (a weight, a numerical one) that orients learning.

[32] Marvin Lee Minsky, who died in 2016, was a US mathematician and scientist specialized in artificial intelligence. Author of *The Society of Mind* (New York, Simon & Schuster, 1986), co-founder of the Artificial Intelligence Project (which later became Artificial Intelligence Laboratory) at MIT and author of numerous texts on AI and philosophy. His positions have influenced controversially the heated debate about the nature and purpose of artificial intelligence.

[33] Arntz, Gregory and Zierahn (2016).

[34] The scenarios of future vehicle automation are described in great detail in Lipson and Kurman (2016).

[35] The story is told in Markoff (2015).

[36] Hendler and Mulvehill (2016), pp. 160–161.

The Datum

Ex-perience Beyond the Experience

I concluded that human action is drawn by the future [...] Hence, the world is full of intelligent anticipators of all shape, dimension and time horizon. We are emphasizing anticipation, but don't intelligent anticipators need past experience to anticipate well? So why say that the mind is forward-looking if it must look backward for its evidence?
M. Seligman et alii, *Homo Prospectus*

It's the end of experience

Since the arrival over the last twenty years of the so-called "experience economy," companies, organizations, markets and businesses have been paying increasing attention to the *customer experience* (CX), demonstrated by their interest in imagining, designing and constructing experiences that can attract, involve, and, perhaps, turn their consumers, employees or citizens into advocates. Evolving from the reflection of Don Norman who coined the expression *user experience* (UX) in the mid-1990s, to the marketing strategies proposed in *The Experience Economy*[1] to the current leading role of the so-called *experience* and *service design* companies,[2] the concept of experience has become a crucial dimension for business innovation, value creation and organizational involvement. Experience is a renewed relational dimension upon which to build services and products (via UX and UI design),[3] branding strategies, marketing and communication (thanks to experiential marketing), organization policies and work, management (through the construction of an employee experience). Today, strategy, planning and delivery of a sophisticated user experience are considered fundamental to the success of a product or service or for memorable branding activities or, again, for engaging the workforce in organizational initiatives. Saying "experience", of course, is to say everything and nothing, a buzzword typical of concepts used today, in very approximate and simplistic ways. For the marketing and communication manager, customer experience is the sum of all the interactions that occur, over time, and through the many points or moments of relational contact between a consumer, a brand, a service, or whatever product of the company. For McKinsey analysts,[4] the consumer experience is one of the most important strategic assets that a company can put in place to differentiate itself from its competitors and create value for the benefit of its customers and consequently for its stakeholders.

While everyone struggles to pursue this Holy Grail, the technological transformations that we are describing (sensors, data, algorithms, artificial intelligence) demand, in my opinion, a radical rethinking of the concept of experience and of the parameters whereby it is imagined and constructed. This new technological apparatus is not only expanding our senses and allowing increased experiences, as we commonly say. In reality, it is introducing, progressively and inadvertently, a new concept of experience. The philosophical effort to be made, then, lies in the rethinking of experience as such. Rather than imagining the extension of the senses or invoking the notion of the "augmented human" (which, of course, will have to be done), we must make another step simply by expanding the concept of experience itself. We consider then not only the increase of the senses, but also the expansion of the experience.

This new expanded perspective, I define with a neologism, *ex-perience*: experience beyond experience.

Human experience is, today, in deep crisis and transformation due to the complex and intricate insertion of the human within the digital and artificial network technologies that operate – and this is the key point – more and more separately and away from the human ways of awareness such as attention, perception and consciousness. And paradoxically, while we try to replicate our sensorial dynamics in machines, machines are progressively moving us away from those processes. Even as we claim that machines have no consciousness, they are evolving in such a way as to make it irrelevant.

We will have to deal with this new "post-experiential" paradigm as philosophers, as citizens and as consumers. Even though we certainly still experience the world through attention, awareness and perception, our ability to access data through digital and artificial networked technologies reformats that experience. It is access to a new sensory reality that brings human experience – as we shall see – to a new and different level of subjectivity and awareness of our being in the world.

This experience becomes multi-scale (regulated by multiple levels), distributed (shared with other subjects), and proleptic (oriented to an anticipated future). A human sensitivity that requires, now more than ever, quantification (data), and that is increasingly connoted – as the digital philosopher Mark Hansen says in *Feed-Forward* – as non-anthropocentric, non-phenomenological, non-prosthetic. In this new perspective, sensory data and cognitive processing go from being a direct access to sensibility (I use my senses to perceive the world) to being mediated by machines (I use data to perceive the world).

If this is what is happening to us, then we must recognize this new sensitivity as "non-perceptual": as the sensing and mining technologies are able to capture new data on the world and orient our future with these data before it reaches our conscious perception, these technologies reposition the human on the basis of an operation on the world that no longer sees it at the center. We increasingly rely on forms of computational mediation that give access and give body to a new sensibility created through data.

Hansen calls *datasense* this new sensitivity that comes with and from data through the use of artificial software-based technologies.[5] Being based on data processing, this new type of experience brings an intentional relationship to sensitivity (the fact that such data are related to our experience of the world) together with an unintentional one (the fact that such data "constitute" the sensitivity). We are beginning to experience a double dimension: data as "methods of access to the sensate world" and data as an effective "new form of sensitivity". In short, the act of accessing the sensate world produces new data on sensitivity and new sensitivity in itself.

If all this is true, it is necessary to quickly understand and evaluate a conception of subjectivity and of agency as radically environmental. We will introduce, in this sense, a new dimension called *elemental* that can account – as we shall see – for the concept of *ex-perience*.

Massive data related to human physiology (heartbeat, galvanic skin response, brain wave activity) as well as those related to environmental and social factors and ecological dynamics (temperature, light, humidity) lead to a shift from a philosophy centered on agency (*agent-centered*) to a philosophy centered on the environment (*environment-centered*). In a speculative conception of the ubiquitous media, not only the screens and networks that animate them, but all matter become an environment of media contagion. There are neither humans nor channels in this elemental perspective where the media are neither fashioned to meet human experiential skills nor humanized, being indifferent and imperceptible to the perceptual and cognitive nature of man (sensation, perception, attention).

In short: sensation without perception, perception without attention, attention without subject. All thanks to data, sensors, algorithms. And, more and more in the future, to the other artificial intelligences that will be developed and discovered. This reconfiguration of experience has very significant consequences for how we conceive of some primary philosophical categories such as, for example, those of time, space and subject. To better understand the new condition, we will look at these three key dimensions and then arrive at a synthesis that best expresses the philosophical transition to post-experience.

New digital temporalities

Speed, acceleration, compression and real time are the concepts that commonly express the relationship between time and digital. Today many people share the feeling that new technologies are impacting our understanding of time in a profound and meaningful way. Our age – so says Manuel Castells – is that of "timeless time", of the shock of an "eternal now" (the *always now* of Douglas Rushkoff), of the domain of "real-time" (the *real-time tyranny*, according to Paul Virilio).

These current paradigms relating to digital temporality are, however – in my opinion – insufficient to account for the profound ontological relationship that data, software and algorithms maintain with the dimension of time. These are all perspectives that do not allow us to appreciate, in depth, the relationship between the dimension of time and new digital and artificial technologies. We must then try to bypass the metaphors of collapse, compression or cancellation of time regarding the size or duration of the present moment, of real-time adherence and acceleration, because they do not help to understand the innovations that 21st century technologies are bringing to the temporal dimension.

The general framework of these traditional approaches is that digitality destroys temporality by accelerating it, compressing it, nullifying it. Our analysis criticizes these current approaches, advancing a more generative and non-destructive prospect of the relationship between code and time. There are two main vectors of this new orientation: *a*) the digital operability of the time and *b*) the proleptic dimension of temporality.

Hansen speaks, philosophically, of the "operational present of sensibility". In short: working on temporal dimensions not perceived by the human, digital and artificial technologies capture in advance the data of a present (time) inaccessible to us and use them to build a future (time) present that will be accessible only subsequently to our perception, experience and consciousness.

In fact, what our consciousness experiences, in reality, as "its" present is already past: the present of consciousness is different from the technological time of sensibility. The present time of the senses is not the same as the present time of the sensors and data. The present of the human is not the present of the machine. "Today's computational microsensors inaugurate the operation of a new level of presencing – the direct presencing of causal efficacy itself (the operational present of sensibility) – that both supplements and, in a sense, take over the long-standing role and privilege consciousness has historically exercised as agent of presencing."[6]

We are sensing a present that, literally, we cannot perceive, to engineer an experience to be lived in the near future and, therefore, to build a present that we will be able to live in. This act opens the experience to an expanded domain of sensibility that has always existed, which has always had an impact on our experience, but which had remained largely opaque to our understanding because it was inaccessible to our (human) way of perceiving, feeling and experiencing the world.

In this new perspective, time becomes a significant dimension in the process of value extraction that is effected via a system that exploits the gap between micro-temporality, machine sensitivity and human consciousness. And therefore, the temporal difference between the present of the machines and the present of the humans generates a new sensibility and new opportunities for knowledge. In this perspective – Hansen continues – the famous "missing half-second" lost between brain activation and the awareness of an event or of a change of condition in one's immediate context is removed and formed without the physiology and neurology of the brain by the digital and artificial technological process. It becomes

the gap between the moment (micro-temporality) in which the technologies mediate/produce new forms of sensitivity derived from the world and
the moment (another micro-temporality) in which they present them, in
an anticipated way and with a process that is defined as "feed-forward",
to the consciousness of the human. What has always been a *neuronal delay*
between our senses and consciousness is transformed, with new technologies, explains Hansen, in a *technical* delay that can be modulated and
malleable in the future.

It is important to point out in a philosophical way that a profound
change of paradigm is taking place: the capture of new sensibilities and
perceptions of the world takes place with temporal dynamics characterized by a fundamental orientation towards the future. We could say that
no longer in "real-time", we now live in "near-time". Rather than mark-
ing the correlation essential between our experience and what is past (the
just-passed), this technically intercepted and modulated delay redirects
everything towards a future moment that will become present. In this anticipation, the operational experience becomes, or rather, will then become
available to consciousness and its future operations (conscious reflection,
decision and, therefore, behavior modification).

The notion of feed-forward becomes central: since traditional perceptual consciousness is simply left out of the loop when technologies perceive and analyze the operational present of sensing at temporal levels
from which conscious human activity is excluded, this operational present can be made available to consciousness only in an immediate future
time. Being presented to consciousness "after the fact", as compared to the
"operational" present, this human present can only come to consciousness
immediately after.

In this perspective, consciousness must be reconceptualized to function in the new post-experiential dimensions: no longer at the center of
the present of sensation, consciousness can be impacted by the current
operationality of the present of sensation only indirectly and in a proleptic
way. The worldly data processed by sensors and algorithms leads to consciousness, not – as we say trivially – in real time, but in an anticipated
contextual micro-temporality.

This is exactly what is meant by the feed-forward structure of the 21st
century consciousness. It grounds what Sadin has defined as the "society
of anticipation"[7] and that Hiu defines as "the era of the third protention",[8]
(projection or anticipation) built through data, sensing technologies, min-

ing and algorithms. So far, historically we have known a "primary protention", that is, the perceived expectation of the imminent moment that
is about to happen, and a "secondary protention", that is to say the imag-
inative expectation based on past experiences. Now, through data and algorithms, we enter into a new modulation of time, a "tertiary protention",
that is, the anticipation of experience built up by algorithmic technologies
and data. In short: temporality and digitality must be seen in a generative,
multiplicative and modulative relationship, and not as cancelling or anni-
hilating time, or merely confused with real, present or accelerated time.

To make a comparison with medical thought, it is no coincidence that
in recent years, against any mechanistic and deterministic reductionism,
even the prospect of an anticipatory medicine[9] is becoming progressively
more relevant. The immune system is an anticipatory system, so is the
female body that prepares itself for pregnancy in advance. Rather than
continuing with the current reactive and remedial medical paradigm (on
the past–present axis), anticipation-based medicine proposes to adopt a
perspective oriented to the future states of our health (from the future to
the present). An early warning system is one whose current state is influenced not only by its past state, but above all by the pace and space of its
future possibilities. And the concept of feed-forward, rather than that of
feed-back, is closely linked to systems of an anticipatory nature. In a feedback process, a system must deviate from the expected behavior to exercise
control; in feed-forward, the expected behavior is constantly modulated on
the future without waiting for a post facto reactive control.

New technologies engineer this proleptic and anticipatory orientation
of our consciousness. In a more general sense, there is a real "precognitive
vocation" of 21st century technologies, a push to the anticipated time that
is effected by sensors and new data processing technologies instructed by
artificial intelligence. The emerging trend that promotes the design of
experiences, the so-called *anticipatory design*, also goes in this direction.
It is the possibility of creating an environment (milieu) that constantly
and fluidly proposes anticipated experiences on the basis of preferences,
profiles and the planning of consumers.

As a first example, let us take *smart objects*. According to the "phase
media" theory as envisioned by Ash, potential multiple temporal dimensions emerge from the activation of intelligent devices and, in particular,
from their capacity to perturb an object or a set of objects around them.
Rather than being directed by a single external linear temporality (past/

present/future), smart objects have the power to instantiate and produce multiple, locally triggered presents, pasts and futures. They create and manage different speed experiences. Think about the smart speakers we are progressive introducing in our lives. When we use Amazon Echo calling Alexa, as Ash points out, a number of perturbations (some faster, some slower) occur between us and the device as well as between the device and the technical environment (router, cloud, network) that support its services modulating, from time to time, slow-down or accelerated temporality experiences.

As a second example, we can look at the timing possibilities imaginable with the arrival of blockchain and smart contract technology in particular. Smart contracts are «state-machine changes», they are contracts that contain certifications, agreements or dispositions that are loaded on the blockchain to start the automated execution. In smart contracts, it is valuable to design and incorporate within the blockchain code the temporality that best meets the needs of the transaction in progress (a lease contract, a notary registration, a monetary lending), assisted by the creation and design of different temporalities. In this case, time can become a "malleable resource" within the executable contracts by algorithmically imagining – says Melanie Swan, a blockchain philosopher – a new time, that is, "blocktime".[10]

Blocktime is different from human time and corresponds to the "time necessary for a certain number of blocks in the chain to be confirmed".[11] In this way, a smart contract could incorporate a slow time (*time slow-down*), an accelerated time (*time speed-up*), a temporality linked to the occurrence of an event (*time event-waiting*) or, still more, a temporality linked to the postulation or assumption of an event (*time event-positing*). Time thus becomes a design dimension (function, functionality) of the blockchain and is no longer specified in minutes or seconds, but in units confirming the blocks of a transaction. It would become a *programmable time* designed and put to work by an anticipatory computational architecture.

In designing these trajectories of different temporalities (we might call it *time design*), there is the possibility of "creating more time" and not, as the old paradigms say, of cancelling or destroying temporality. With blockchain technologies and the progressive introduction and adoption, for example, of smart contracts, the creative potential of machine learning and big data becomes exponential, especially in their ability to become builders of these future advances of an "economy of automation" (and not anymore focused, as it is still today, but to an ever-lesser extent, on the consciousness and experience of the "economy of attention").

The birth of code/space

As it occurred for time, the spatial dimension arising from its relationship with digitality and coding has been and continues more often than not to be dominated by destruction and elimination concepts. People speak of the compression and elimination of space, of the end of geography and distances, of the irrelevance of the body and of materiality. These obsolete paradigms still occupy the current thinking about the relationship between code and space. In this sense, the use of the term "virtual" has coincided in the past – and in part still today – with the idea that space and code are in a mutually exclusive relationship. On the one hand the geometric, material and concrete space, on the other hand the digital, immaterial and virtual technologies. The early season of successes (and failures) of the first «virtual worlds» all the way to Second Life, has coincided with this devaluation of the spatial, geographic and corporeal dimension. How far from the truth these connotations are today is underscored by the success of virtual reality viewers in physiotherapy to rehabilitate patients seriously injured in their motor skills.

With the massive diffusion of mobile technologies, from the internet of things to the industrial internet and robotics, what we are witnessing is, instead, a significant return to consider the relevance of the dimension of space and mobility and its deep relationship with digitality. In particular, in the last decade, studies on *mobile media*[12] have multiplied together with the analysis of *new mobiliti*es.[13] Furthermore, the impact and diffusion of mobile devices and media, autonomous vehicles, smart homes and cars have given new strength and substance to a digital thinking in which space and the environment have once again become relevant. Almost thirty years ago (in 1991, to be precise) Mark Weiser wrote prophetically that we would completely reverse the prospect: from bringing the world into computers (by duplicating the world within websites and applications) to bring the computers into the world (i.e. incorporating computational skills into objects and environments).

Finally, the unexpected revival of augmented reality (AR) and also of virtual reality (VR) thanks to the driving force given by Facebook with

Oculus Rift (and today many other brands), and the recent enrichment of VR visors' functionalities have definitively established this "return to space" (*spatial turn*). This is obviously a return to a very special spatiality because what we are talking about is not the space we knew before the arrival of software, code, data, sensors and artificial intelligence. It has gotten to the point where we have been forced to update the dictionary with various neologisms: "hybrid spaces", "mixed reality", "cyber-physical systems", "phygital world". New labels are proliferating, but – in my opinion – they are not enough. Even the well-known Castells concept of the network as a "space of flows" does not help to understand the true relationship between spatiality and digitality. They are not sufficient because, fundamentally, they fail to explain the ontogenetic relation that exists between space and code.

On the contrary, digital geographers and urban planners are rethinking the concept of space and its ontology in the light of code that is worn, transported, incorporated into objects, movements and environments. Considered as a fact, as an absolute reality (metaphorically: a Cartesian geometric cube within which Galilean objects move), over time, space has instead assumed the characteristics of a relational, social, contextually built concept.

These characteristics have been, in some way, accentuated and taken to their extreme, in the last fifteen years, by those who reasoned not on what space is, but, precisely, on how it becomes – thanks to the code – that which it is. In this perspective, space is therefore a constantly evolving production, and software code is the key element that continually produces space and its meaning. This generative operation of space activated by the code (sensors, data and algorithms) is called *transduction*.

This concept underlines the capacity of software to produce the conditions of existence of a kind of space, constantly modulated and iterated in its relations of proximity, movement, sociality, sensitivity, accessibility and so on. Spatiality is the result of this codified production. This is not – as people often say – a case of fusion between physical space and digital services. No fusion is in progress. Continuing to use this metaphor does not help us understand.

Rather, space is "transduced" in forms and modalities that are emerging, at the same time contingent and contextual. The "code/space" is the concept[14] that makes explicit the new condition of a spatiality that is the product of the generative capacities of the code, of sensors, data, algorithms and artificial intelligence that do not cancel it, but multiply it exponentially.

Let's give two examples to make this philosophical perspective more explicit. The spatiality of an airport check-in area depends entirely on the software. If the software fails, that spatiality stops functioning as it was designed to and how it should; there is no way to proceed manually or mechanically as the procedures were all computerized and coded in the systems and software programs responsible for the check-in activities.

The same thing happens to a supermarket where today the information processes (from the warehouse to the cash desk) are computerized. If the software stops working, for example at the cash register, that particular space loses its meaning: at the instant the code collapses, a supermarket loses, in theory and in practice, its spatiality as such and becomes a simple warehouse. Its spatial ontology as a supermarket disappears only to return when the software re-establishes its spatial nature as a supermarket. Or, put in another way, software can completely eliminate cash and queues. The designers of Amazon Go, the new physical store that the American giant has very recently launched, have perfectly understood the transductive potential of the code/space. There are no files or cash in the new store. You enter by authenticating yourself with your smartphone, you choose products from the shelves and, once put in the bag, you simply leave. Computer vision, deep learning algorithms and sensors are responsible for invisibly and automatically recording the purchases made, for making payments and for sending mobile and digital receipts.

In some cases, code and space are not totally integrated, so there is still the possibility of making the space work, albeit in a less efficient way, if the code doesn't. Or, in another case (as we saw in The Sensor chapter), there is the kind of medical code that, embedded in a house, will be able to transform it into a personalized hospital unit. The same applies to intelligent objects or, as they are called, *codjects* (code-objects): their ontology increasingly depends on the code, data and intelligence implanted. The code decides what that space will become, its ontogenesis.

This generative link between code and space is more pronounced and evident in contexts such as VR and AR. The digital image, be it in the form of VR or AR, in fact passes from being a purely geometric projection to being an algorithmic process. In these cases, reality is no longer a fixed representation of the world, but the programmable "instantiation" of a database in real time. As Mackenzie wrote, we will have to start speaking of *dataspace*.[15] In these cases, "the image is nothing but the moment of network access"[16] to its sensors, data and intelligent algorithms (people talk about *softimage*, a contraction between software and image). In grammatical terms, you could say that the status of the image evolves from being a mere noun to becoming a verb, that is endowed with a power of performance.

Once again it is opportune to enter philosophically into this new ontology. The virtual or augmented image is not what has been recorded or archived once and for all, but what is temporarily and each time performed. It does not exist as an archive of those images. Digital and artificial technologies do not re-establish a photographic past, but rather instantiate a present signal.[17] Virtual or augmented reality does not function as a mere representation: it is the actualization of data in the network with access, transfer and visualization, removing the boundaries between "visual data" and "visualization of data".

Although our experience and visual perception are still linked to the photographic paradigm of images, digital spatial images have a different ontology and follow the new algorithmic paradigm. We are thus moving from geometry to the algorithm – according to Hoelzl and Marie – and from the geometric projection (which was the operating mode we have used for a long time) to the algorithmic processing (the algorithmic image governed by databases and networks). The latter also changes in relation to the different temporality that is triggered because – as we said – the image is not archived from past experience, but instantiated in the present. The archive, here, does not serve to preserve (representation), but to reactivate (operational). As we have learned, as a creature of the code, this reality is not conceived as a registration of an impossible past, but as an actualization of a possible future through data.

Elemental, my dear Watson!

When we talk about digital and artificial technologies on the net, as we can guess, the themes of "subjectivity" connected to identity (who is what), to sensitivity (who perceives what), to agency (who does what) and, naturally, accountability (who is responsible for what) are crucial. Philosophy, technology and jurisprudence contend the relevance of these dimensions, which in turn are central to the ways whereby, eventually, we can create value for consumers and citizens and develop the ethical governance of public affairs and private business.

Intelligent objects, assistive bots, coded algorithms, anticipating software, autonomous vehicles, quantified bodies, data-driven agents, all demand of us, therefore, to address the philosophical question of the subject with new perspectives. But what kind of subjectivity is emerging?

The foundational idea that we explore here is that of the subject considered as an "elemental" entity. Elemental is a term used in media studies to indicate the new dimension of presence and action introduced by sensory, digital and artificial networks and their dynamics operating at levels and scales which are, at the same time, above and below the human sensitivity. We must not, of course, confuse elemental with environmental understood in a naturalistic sense. In an elemental perspective, subjectivity must be rethought and reconceptualized as it is no longer identifiable as the privileged prerogative of individual human actors.

This philosophical perspective, therefore, insists on the non-human aspects of the network dimension, orienting itself to embrace a dispersive and distributive dimension of the capacity to act (agency) within the networks of sensors, processing, implementation and communication. Therefore, a fundamental rethinking of the human and of human experience is in order to be able to fit them adequately into an elemental thought. We must also abandon the easy dichotomies between human and non-human and begin to think of a human as one of the possible declinations or embodiments of elemental subjectivity. And, then, adopt a perspective where subjectivity is an emerging configuration of the elemental between the micro-scale (sensors, actuators, transducers, perturbations) and the macro-scale (networks, clouds, protocols, platforms).

In so doing, we must, for a moment, withhold the idea of agency and the notion of smartness as connected and similar to the common understanding of human intelligence: computer science and moral philosophy are not on the same wavelength when they talk about agency. We must open ourselves to considering entities that act in the world, but without necessarily having the characteristics that, as a rule, we assign to human intentional action.

We can now try to indicate some of the main characteristics of this new "elemental subjectivity"; although they are separate characteristics, they are strongly intertwined with each other.

Data-driven agency

The first important aspect is that we are dealing with a kind of *data-driven* subjectivity (human as well as non-human) that is guided or enhanced by data, from the *data-sense* we mentioned at the beginning. Hansen talks about *data-enhanced actors*. Datasense implies the production of sensitivity due to data collection operations. To be clearer: the data create a totally new domain of sensitivity (*data as experience*). The perspective is dual: access to data is also data production. The data do not mediate (if mediate is still the right verb to use) our senses, but mediate sensation as such and in itself. Sensibility mediation is also sensitivity production.

The core of this data-driven agency will increasingly be *machine learning* and its ability to model reality in a predictive advance for the future. And if a machine or a data-driven agent defines a situation as real, it is also real in its consequences. So, regardless of whether or not we decide to consider real and concrete (and not just immaterial and virtual), algorithms, bots, assistants and artificial agents, they are certainly real in the consequences of their actions. I also started to rename as "infoviduality" and "infovidual" this emerging data-driven agency to stress the protentive – not only retentive – dimension of computing and modeling data (biometric, sociometric, ecometric, etc.). In our perspective, retentive means past-oriented data and protentive means future-oriented data.

Distributed agency

We also talk about subjectivities that are distributed. Subjectivity is distributed between micro and macro dimensions, between scales and multiple levels, between probability calculations and forecasting models. In fact, what is happening is that the philosophical, religious, historical and cultural systems that have given life in the past to the concept of the "individual" (and to its counterparts such as community and society) are eroding in favour of an emerging age of "dividuality"[18] – as the philosopher Deleuze would say – with all its impacts both in a positive (expected empowerment) and also in a critical perspective (risk of enslavement). In fact, in the calculation of the "human", we have moved from the age of the average individual to the age of digital dividuality and now to the age of infoviduality.

For this transition[19] from "individuality" to "dividuality" to my concept of "infoviduality", the calculation and quantification of the body and its organs, of gestures and movements, of behaviors and interactions become the means to construct the idea and the practice of a new subjectivity, multiplied, distributed and mostly anticipated. Objects and bots animate interactions the responsibility of which is often not easy to place and determine, spread as it is between algorithms, networks, databases and protocols (not just humans). We are used to say that media are "extensions" of humans. Instead, I think it's time to say that, in many cases, media are and will be "abstentions" of humans.

Automated agency

The elemental subjectivity is, moreover, more and more automated, whether it concerns human subjects or refers to non-human entities. In computational worlds, artificial agents are a type of software able to act autonomously, perceiving in some way the external environment and producing a certain adaptive change. A car without a driver and a self-regulating thermostat are active, an energy network that distributes services using real-time feedback on consumption to determine prices and disbursements is also an agent. There are, of course, different levels in the ability to act:[20] agents driven by deterministic algorithms (perception and response are predefined), agents that employ learning algorithms (and that, in a supervised, unsupervised or reinforced way, build new knowledge), agents based on a multi-agency system (entities of the first and second order that interact even, on occasion negotiating their respective objectives in an emerging way), or completely autonomous agents (biological or artificial, able to survive outside the software architectures even if they are constituted by computational systems).

But automation is not just related to robots or machines. I think automation is becoming a more general orientation of our society. I tend to consider automation, paradoxically, as a "new (human) institution". To me automation, to give an example, is also instantiated by the blockchain protocol: in this case, the protocol is an automated algorithmic mechanism that mobilize self-enforced interactions among nodes in a peer-to-peer network.

Precognitive agency

Finally, there are subjectivities anticipated by the feed-forward structure that we have analyzed above (anticipatory action, tertiary protection, near time). If the media and mediation technologies of the 20th century had operations based on registration, archiving and transmission, the sensing and mining technologies of the 21st century are a new architecture that enhances the new anticipated temporal dimensions). I started to frame metaphorically this paradigm change as the shift from media technology as "archive" to media technology as "oracle". We no longer live in an archival age, we now live in the oracular age. As Mireille Hildebrandt, the philosopher of the relationship between law and technology, says, "Pre-emptive smart environments begin to transform our dealings with artefacts. At some points, we will become aware of the fact that we are being watched and anticipated by machines and we will try to figure out how the infrastructure 'reads' us and with whom it shares its knowledge of our preferences and of the risks we incorporate."[21]

Going from the analysis of profiling a behaviour to that of anticipating it would seem to be a brief and innocuous step, but in reality, it has deep epistemological and ontological implications. Philosophically, we must ask ourselves: is this an *anticipation* of the future (let's say, for example, that we would have bought such a book in the following months anyway and that the algorithm was only anticipating our action) or it is about the actual *creation* of that future (in fact we would never have discovered that book and would never have bought it if it had not been suggested by the platform)?

Then how do we distinguish between anticipation and creation? In the case of crime prevention activities, would the repeat offender have or have not committed an offense if we had not anticipated it and stopped it with the help of modern precognitive technologies for predicting criminal offenders? This brings to mind the precognitive investigations of Tom Cruise in the film *Minority Report*.

There is something more, however, and we should allow a small digression on the relationship between law and code. Let us ask ourselves: what happens to the mode of existence of the law with the emergence of digital and artificial technologies.

The title of Hildebrandt's essay could not be more explicit: *Smart Technologies and the End(s) of Law*. What would this end of the law caused by smart technologies consist of? The law as we know it until now has typical characteristics: *a)* the written law highlights the norm and makes it independent of its creator and interpreter; *b)* the distance between law and its creator/applicator implies that interpretation becomes the characteristic feature of the law; *c)* the extracted norm is the foundation of wider and different political and institutional applications; *d*) there is a class of interpretations that serve as buffer between the legislator and the subjects; *e*) thanks to the press, the written law is strengthened, diversified and democratized; *f*) the proliferation of the printed law implies the proliferation of subjects that reinterpret the interpretation; *g*) the role of law emerges as a shared normative element in need of interpretation.

Now let's look at what happens with new technologies – explains Hildebrandt. In algorithmic society, there is a gradual incorporation of the law in the software code. Behaviors and rules are increasingly grafted into devices, applications and platforms, being the result of the design of technicians, designers and managers: there is no legislator. It becomes increasingly difficult, once you accept the configurations and biases of the platforms, to violate or disobey them: no device builder embeds in its application mechanisms to allow it to be violated, whereas traditional laws cannot prevent disobedience. The law incorporated in the devices is more stringent. The contestation of technologically grounded laws and regulations is difficult because most often the technological mechanisms and dynamics – as we have learned with code, data, and algorithms – are invisible. The way of the existence of the law, as we have built it and known it up to now, is connected to the printed paper and thus has configured our world. The printed paper and the texts form the preconditions for the diffusion, comprehension, interpretation and implementation of the laws and regulations. It is no longer like that with a code law. So it is clear that this new "technological incorporation" has to be philosophically investigated.

The rules encoded in the software are built into an invisible complexity (therefore difficult to know and to criticize); the interpretative distance is ignored by the performance of the machine and the algorithm; the libertarian freedom connected to technologies can produce phenomena of anarchy in which new market monopolies or liberticidal institutions proliferate; the buffer role of the interpreters is henceforth played by engineers and programmers who, generally, are neither neutral nor legally competent; it favours informed people, but with a quantitative, cybernetic or computer science footprint unfamiliar with other more interpretative paradigms (computational thinking tends to exclude ambiguity and to ensure control only by data-driven feedback mechanisms); finally, no element or moment of suspension of the judgment is foreseen. Everything happens mechanically, algorithmically and automatically (see, for example, emerging smart contract practice on blockchain protocol).

More generally, in fact, we are witnessing a double ontogenetic movement: the code is incorporated and instrumented in objects and environments; vice versa and at the same time, the logic and dynamics of the
assignment of "identity", "existence", "transparency", "trust", "law" and so
on are incorporated into the elemental code – so to speak. It is not a matter of little consideration, as one can imagine. Now, let's get back to the
elemental code.

Data as ultimate interface

Now that we think of it, many of the metaphors that software engineers,
data scientists, and platform architects use have to do with atmospheric
phenomena: *cloud infrastructure,*[22] *smart dust,*[23] *fog computing.*[24] Almost
unconsciously, it would seem that experts in technology and innovation
have begun to perceive this shift in the elemental paradigm that we have
tried to describe so far, to the point that people are beginning to talk about
atmospheric media.

As Hansen writes:

> Exemplified by networks, atmospheric media operate through the radical
> technical distribution and multi-scalar dispersal of agency. Simply put: sub-
> jectivity must be conceptualized as intrinsic to the sensory affordances that
> inhere in today's networks and media environments [...] In our interactions
> with the atmospheric media of the twenty-first century, we can no longer con-
> ceive of ourselves as separate and quasi-autonomous subjects, facing off against
> distinct media objects; rather, we are ourselves composed as subjects through
> the operation of a host of multi-scalar processes, some of which seem more
> 'embodied' (like neural processing), and others more 'enworlded' (like rhyth-
> mic synchronization with material events).[25]

We therefore need a neutral theory of experience, which applies to humans
and non-humans and which breaks the separation between the animate
and the non-animate. Once upon a time there was the experience – one
might say – an *ex-perience,* precisely. Today, experience cannot be the privilege of a single class of actors (the human ones), but it must be reconceptualized to the point where it can include a wide range of sensory and
cognitive possibilities. Experience involving the human, object-centered
and body-centric needs to be upgraded.

Since the impact of new technologies occurs in a peripheral and distributed, under-perceived and/or vice-versa happening at the level of macro-scale and networks, we can no longer assign the privilege of subjectivity to perception and consciousness. Shaviro speaks explicitly of *discognition* and of other kinds of intelligence that we should be aware of: "Sentience, whether in human beings, in animals, in other sorts of organisms, or in artificial entities, is less a matter of cognition than it is one of what I have ventured to call discognition. I use this neologism to designate something that disrupts cognition, exceeds the limits of cognition, but also subtends cognition."[26]

Perception and consciousness will increasingly become elemental dimensions. Artificial sensing technologies will directly mediate the causal infrastructure of worldly sensibility in micro-temporality and micro-sensoriality independent of consciousness, fueling our near future, influencing the action of our consciousness in the world and shaping the active role of the person aimed at achieving their goals (feed-forward) or the goals of those who will have the practical and political opportunity to act on these dimensions. In fact, Hansen cautiously and correctly questions: "Will the atmospheric media systems of the future provide opportunity for open-ended intensification of our experience, or will they remain exclusively focused on instrumentally targeting specific effects aimed at making our 'desire' legible for exploitation by others."[27] In any case, every act of accessing sensitivity data will itself be a process that creates a new sensibility. As I happened to say: the data is and will be our ultimate interface with the world.

Data mining technologies and processes do not simply calculate a pre-existing space of possibilities, but literally create new information as a result of the operations implemented. In this perspective, moving from media to market, the data represent a potential paradigm shift in economic institution. It's not just the mantra that "data is the new oil", but that data could be "the new money". In fact, as Viktor Mayer-Schönberger and Thomas Ramge point out in their recent *Reinventing Capitalism in the Age of Big Data*, money-based markets are more and more obsolete and inadequate transactional technologies. In data-rich markets, instead, we no longer have to condense the people preferences information into prices (using finance and money as technology to create, coordinate and manage business and organizations transactions). We can use data flows, information richness and machine algorithms to redesign and create more efficient and innovative markets and organizations. If this is true, we are just at the beginning of a new economic revolution based on data.

It may not be easy or painless, but I guess the rough direction is marked. Not entirely and precisely determined, but oriented and traced in a way, it seems to me, that is sufficiently clear.

It will not be easy. To give an example – following Pentland[28] – our identity is still today fundamentally an analogical question within a world that, in the meantime, has become digital and artificial. We should therefore begin to rethink, through the data, the identity or, better, digital identities that, on our behalf, will make new experiences of the world (and the many related concepts: our data, trust mechanisms, forms of agency, collective responsibilities, etc.) in terms always more digital, networked and artificial. The World Economic Forum also acknowledged this in a recent document on the subject.[29]

Given the importance it will have, I think it is our duty to be aware of the sociocultural dimension of data.[30] The awareness now growing in research environments is that data is not just "given", but that it is the concrete historical result of the specific cultural, social, technical and economic choices that are put in place by individuals, institutions or companies to collect, analyze and use information and knowledge. Perhaps less so in practices of the market and of companies. The social construction of data is not a new perspective: the concept of *raw data* is an oxymoron – say philosophers, anthropologists and sociologists.[31] The so-called "raw data" (that is to say data not contaminated by theory or analysis or context) does not exist, but is always the result of a data journey made of operations and elaborations of various kinds, also very material in their composition (production, processing, distribution and so on). We talk about the materiality of data[32] from many perspectives, including political ones. As Dourish suggests, the materiality of the digital datum lies, first of all, in being an information that as a material-semiotic entity is physically stored in disks and magnetic supports and/or that is transmitted wirelessly through electromagnetic signals. The data have dimensionality, weight and structure. In addition, they are the product of a set of practices, technologies and material devices. On the other hand, they are also material in their consequences. Materiality is expressed, therefore, in the conditions of production from collection to archiving, in the physicality of support infrastructures such as computers and processors, in concrete properties such as their durability, dimension and mutability in time and space, in conditions that impact on environments, processes and operations.

Data is the key element of our time: the new oil, the new energy – they say. Even more radically, data is the new money, say others. The data – we have seen – feed code, algorithms and artificial intelligence. But there is another large consumer of data that still only a few people are aware of today. It uses data to govern, even politically, a world that is increasingly configured as a platform of platforms. It is time to get to know "The Stack".

Notes

[1] *The Experience Economy* is the key text by Pine and Gilmore (1999) that describes the characteristics of a new form of economy (after agriculture, industry and services) in which experience plays a fundamental role in the construction of the relationship with consumers and in the generation of value.

[2] As Katz (2015) recalls, Herbert Simon provocatively suggested extending the domain of the new design science beyond the aesthetics of objects. From the original provocation of Simon, *design thinking*, as an idea of applying the designer's tools to the totality of life, has actually gained consensus. In Simon's idea, that radical expansion of the perimeter of design had to lead not only to design functional and aesthetically cared for products, but to represent an approach to the overall design of the whole field of human experience.

[3] UI (*user interface*) design is the set of activities of analysis, design and prototyping of the user interface, while UX (*user experience*) design is the analysis, planning and realization of the user experience. The first focuses on the interface that will need to be developed to create a satisfying user experience, the latter on optimizing the empathetic experience built on the characteristics and needs of the user.

[4] McKinsey (2016). This report not only underlines the importance of imagining and proposing to the consumers performed and memorable experiences, but also emphasizes the importance of reorienting the whole organization with this vision centered on the customer experience.

[5] Hansen (2015) speaks of a sort of new sixth sense that would have in the data its primary cognitive vector and which therefore defines *datasense*. Hansen's perspective recalls a philosophical orientation that is emerging in these last years and that goes, albeit in multiform ways and with different reflections, under the label of *object-oriented ontology* (OOO). To simplify, it is a paradigm of thought that tends to consider objects as entities equal to the human, as capable of existence, autonomy, relationship, communication and interaction in itself. Two primary assumptions: *a*) the non-centrality of human subjectivity and *b*) the independent relationship between objects. Critics and supporters animate the ongoing debate on the principles and perspectives of an OOO.

[6] *Ibidem*, p. 192.

[7] Sadin (2011).

[8] Hui (2016).

[9] Nadin (2016).

[10] Swan (2016). In this text dedicated to the question of modifiable temporality through blockchain technology, Melanie Swan urges the business community to consider the new protocol as a tool to build economic ontologies and new businesses: from the construction of codified contracts (smart contract) to distributed and automated organizations (DAO).

[11] *Ibidem*, p. 191.

[12] On the topic of local and mobile media, it is useful to compare the overall work of Jason Farman, starting with the essay *Mobile Interface Theory*. Another work, *Embodied Space and Locative Media* (New York, Routledge, 2012), investigated the relationship between media and locative and site-centric media.

[13] Studies on the so-called *new mobilities* for about a decade, with the works – among others – by John Urry, have analyzed the new dimensions of geographic and spatial mobility working in an interdisciplinary way between anthropology, sociology, geography, transport and logistics, migration and cultural analysis. See in Sheller (2016) a summary of the state of the art.

[14] The concept is proposed and analyzed by Dodge and Kitchin (2011). More specifically, both digital geographers and urban planners identify three modes of relationship between space and code: 1) the code/space (*code/space*), 2) the codified space (*coded space*) and 3) the space coded in the background (*background coded space*). In the first case, there are processes and economic activities in which the code completely dominates the process of producing space; in such cases, if the software code does not act, the space is not modulated and the connected social relations have no way of being developed. In the second case, the relationship between code and space is not so close and vital and, in fact, if the software is not activated or does not work properly, space and relationships can be modulated, even if in a less efficient and riskier form. Finally, there are cases in which codification processes have the power to produce and mediate social and spatial relations on condition that they are intentionally activated; in fact, in different contexts, the space has incorporated the software code that remains in the background until it is made operational. Once the dormant code is activated, it then assumes one of the two previous forms: vital or supportive.

[15] Mackenzie (2010).

[16] For an in-depth and critical analysis of the ontological status of images in the era of their algorithmic reproducibility, the work of Hoelzl and Rémi (2015) is useful, in whose perspective the images have lost the function of iconic representation and they have become a data-to-data relationship. The image is not only part of a program, but also contains its operating code: the image is a software program in itself.

[17] Hoelzl and Rémi (2015).

[18] The concept of dividuality is linked to the thinking of the French philosopher Gilles Deleuze who uses it to indicate how individuals, in a data-driven society, become an expression of the multiplicity of calculations that are processed on their information.

[19] On the concept of dividuum, we can see Raunig (2016).

[20] Hildebrandt (2015) reviews the multiple forms taken by a data driven activity, applying the lens of the philosophy of law and technology. Together with the themes of the ability to act on artificial agents, related topics such as security and governance, privacy and personal data protection, opacity and transparency of new technologies, rights and individual freedoms in an artificially intelligent world are investigated.

[21] *Ibidem*, p. 11.

[22] In computing with the English term cloud ("cloud computing") we indicate a paradigm for the supply of IT resources, such as storing, processing or transmission of

data, characterized by availability on demand through the internet starting from a set of pre-existing and configurable resources.

[23] *Smart dust* is a hypothetical network made up of microscopic electromechanical systems (*mems*) connected to a wireless system and capable of detecting (for example) light, temperature or vibrations.

[24] *Fog computing* is a new computing paradigm linked to the IoT that provides distributed storage computing to bring the cloud closer to the *edge* of the network, bringing this service as much as possible near the end users.

[25] Hansen (2015), p. 3. This shift towards the ex-perience also has an impact on branding and advertising strategies and practices, in particular on the critical issue of "efficiency loss of the symbolic" in favor of the algorithmic. On this point, the text of Brodmerkel and Carah (2016) is fundamental.

[26] Shaviro (2016), p. 10.

[27] Hansen (2015), pp. 132–133.

[28] Pentland, Shrier, Hardjono and Wladawsky-Berger (2016).

[29] World Economic Forum (2016).

[30] One cannot, of course, study the data regardless of the triad in which, since the 1980s, researchers have inserted it, which consists of: data, information and knowledge. As a rule, these three dimensions are represented as a pyramid, at the base of which we find the data, at the center the information and at the apex the knowledge. In the classic hierarchical relationship (pyramid of knowledge), the data are transformed into information and this distillation into knowledge. It should however be pointed out that the meaning of the three terms does not have a univocal connotation and is, still today, variously interpreted by different approaches and analytical perspectives.

[31] For a critical perspective on the production of data seen through the lens of philosophy, anthropology and sociology we can suggest two key texts: Gitelman (2013) and Boellstorff and Maurer (2015).

[32] Dourish (2017).

The Platform

Emerging Accidental Megastructures

> They form an accidental megastructure called The Stack that is not only a kind of planetary-scale computing system; it is also a new architecture for how we divide up the world into sovereign spaces [...] To be clear, this figure of The Stack does and does not exist as such; it is an idea and a thing; it is a machine that serve as a schema
>
> B.H. Bratton, *The Stack*

Living in a stacked world

The code is an executable writing of the world. It produces action on our life and on our reality. As we have learned, it has the ability to make things happen. If we think about it, this is precisely one of the founding prerogatives of *sovereignty*. And, ultimately, a sign of power. In fact, we started this book talking about taking control of the software. And many questions have emerged about governmentality and agency, individual as well as collective, private and public, human and inhuman, conscious and unconscious, local and scaling up.

Certainly, it is no longer just a matter of the government understood in legal, contractualistic or biopolitical terms,[1] but more and more of the exercise of power in terms of protocols and algorithms. Quite the opposite, increasingly traditional aspects of governance of self and social are being subsumed and reabsorbed by forms of software-driven management and modulation (and control). As Yuill wrote: "Software engineering is simultaneously social engineering."[2] And we must also consider that software is always social:[3] if anything, it is about understanding what kind of sociality, from time to time, it is called to embody. Calling "social software" only the code that underlies social networks and social media – as we commonly tend to think – is a dangerous short-sightedness.

So really at stake are agency (action) and the ability to govern (government). This is particularly evident in the emerging smart contracts on blockchain protocols: legal enforcement is imbedded in the writing of a code that makes the contractual clauses algorithmically executable. More: for some philosophers of the law – such as the aforementioned Hildebrandt – the arrival of digital and artificial technologies is undermining jurisprudence as we have known it so far. However, it is not merely a question of jurisprudence or economics.

Ultimately, in a broad sense, it is a political question. *Software and sovereignty* is precisely the evocative subtitle that Benjamin Bratton gave to his book *The Stack*. In this thick and full-bodied volume, the American digital theorist analyzes, in a visionary and critical way, the emerging impact of the kind of computation that today, increasingly governs on a planetary scale. Computation that is not just a matter of codes, data, sensors, architectures, intelligent machines, but it expands to understand and redesign geophilosophical, geopolitical, geojurisdictional and geoeconomic issues in a world that has become programmable. A computation which, in his opinion, gave rise to an accidental megastructure called "The Stack", acting on multiple levels and dimensions, modulating and rewriting the logics, the dynamics and practices of new kinds of sovereignty in which software plays a key role (sovereignty of the self or individual, but also and at the same time social, economic, cultural and political sovereignty).

Both in its name and in the engineering imaginary, the metaphor recalls the layered architectures – hardware and software – that structure and animate the digital economies: in reality, as Bratton himself says, the stack exists and does not exist at the same time.

Although abstract, it is a philosophical perspective (we could call it a mental experiment) that allows, however, to highlight the different concrete operations, enabled by software and the sensing, processing, networking and actuating technologies. Through these operations, today we define and are defined in our identity, we access goods and services enabled by cloud platforms, States and nations manage exchanges, flows and policies of participation, inclusion and exclusion, the companies and logistics organizations evaluate and procure environmental and human resources for production and commercial purposes.

As with many of the philosophically based analyses that we have had the opportunity to present, the latter also undermines the foundations of many conceptual assumptions, interpretative categories and structures of thought to which we have been educated. At the same time, it encourages us to imagine a possible scenario of synthesis, connecting the various themes that we have touched upon through our speculative path.

The computational paradigm therefore invests multiple entities and processes on a planetary scale: the procurement of energy, mineral resources and the transport network, the cloud infrastructure of architectures and platforms, urban software and the privatization of public services, a massive universal system for identifying and situating people, objects and environments, intelligent interfaces, more or less recognizable as such, able to increase human sensory and cognitive capacity, quantifying subjects tracked by the measurement of legions of sensors, algorithms and machines.

All these (and many other) manifestations of the code rather than being analyzed as separate components are to be understood in that emergent, multi-layer, pervasive architecture that we call the stack. It is a mega cumulative accidental architecture (therefore without any intelligent, dystopic or utopian unitary design) that, due to the technologies we are progressively introducing and adopting, is reconfiguring, in a more or less visible and more or less contradictory way, geographies, economies, politics and cultures. In this perspective, the software theory (and the connected software design) we talked about in the initial chapter, must be articulated nowadays as *platform theory* (and *platform design*).

The stack can in fact be described as a crossover, an interweaving of platforms: as if it were "the platform of platforms", that is to say a complex architecture that intersects the extraction of materials and resources, the activation of infrastructures and logistic flows, the implementation of cloud services and applications, the urban algorithmic regulation and the positioning of objects and people in the network and in space, the continuous predisposition of interfaces (even down to no UI) for the interaction of human and non-human actors.[4] The stack emerges from the stratification and overlapping of several interdependent layers: the earth, the cloud, the city, the address, the interface and the user. Even if it is a question of tracing boundaries between levels that are not always easy to maintain, each layer has its own peculiar characteristics.

In the stack, the computation is physical, made by platforms that have a huge hunger for molecules and atoms, as well as for bits. We started this journey with the preoccupying idea that software was eating the world, now we discover that software in the form of a platform is conquering the planet. In fact, platforms design new sovereignty regimes (*platform sovereignty*), a sovereignty, still immature and emerging, exerted on collective physical and digital dimensions.

In this will to govern, the stack intersects and questions other entities that also claim rights to dominate and control the humans and the planet: political institutions, nations and states, organizations and businesses, local bodies and municipalities, transport networks and logistics.

To give a concrete example of this conquest of "political" sovereignty by the platforms, we take the cartography services, once a sphere of power and control logic of states and nations. And, as cartographers know all too well, he who maps the territory controls it. Today more and more applications and services of mapping and geolocation are prerogatives of private platforms (Google Maps, to be clear). This discourse, however, relates to a kind of geography that is increasingly regulated by the code, as we saw in the initial chapter. But if, in that case, normativity was linked to the ability of the code to transduce space (to represent the space as a continuously codified environment), in the stack normativity is more explicitly highlighted as a governing (political) control over the territory. The Google Maps service that allows us to visualize the space through its digital maps, updated in real time, connected to cloud technologies and social networks, has, in fact, the sovereignty over those movements, on the localization and the positioning of the user in the path, on the interpretation of that space through the information transmitted and displayed on the device. And, lastly, it governs the cognitive modalities that it implicitly conveys via photographic and cartographic ensembles that are, as we know, the "operational images" of *softimage*.

How, then, is the stack articulated?

Earth to cloud and back

As Bratton explains, The Stack is made up of six layers, all strongly interconnected, worth examining in detail in order to highlight their own logic and dynamics.

The earth

This is the layer that provides the geophysical foundation to the stack. It is important both because it supplies the geological substratum for computational hardware, and also because it is closely linked to the geopolitics and geo-economics of mineral resources (extraction, consumption, waste) and to that of flows and vital energy supplies to support the other platforms (e.g. the data center locations of the platforms or the increasing energy consumption required for data management). There is no computation without transforming matter into energy and energy into information, and that is why the relationship between depending on the earth and on its substratum is vital to computation. And the stack is a very hungry machine for planetary energy, minerals, metals and chemicals (our smartphones, batteries, sensors, cables and processors are made of these).

Precisely because the processes of computation are physical events, these appetites collide with the finitude and control of the planet's resources (and not only because of the mathematical limits of computation or the political limits of hierarchies of geographical sovereignty). The stack is not merely on this planet and built from this planet. It is also a way to redesign the planet, transforming the geographic and geopolitical elements present in the territories and delimiting and delineating separate areas through the layer dimensions that compose it (city, interface, cloud). Naturally, while designing new geographies, the stack must negotiate with the existing ones and with the limits imposed by them.

The cloud

This is the layer that manages the entire infrastructure of massive servers and databases, submarine optical cables, satellite technologies, services and distributed applications, producing the ubiquitous computing that governs the stack. It is the level that, overlapping and juxtaposing itself with other jurisdictional entities, further questions borders, domains and geopolitical sovereignties. The conflict that opposes China to a platform company like Google is a concrete example, as is the comparison of Facebook vs Europe on the data management of WhatsApp or the clash between Airbnb and the regulatory bodies of the municipalities of New York, San Francisco, London, or Amsterdam. This is not just a question of the presence of the American search engine and its economic services in China or of a dispute over the protection and regulation of personal data between the USA and Europe. More speculatively it is the clash for dominance, control and power of two different ways of deploying sovereignty: the old nation-state against the new platform model supported by cloud technologies. The sphere of activity of the cloud is, in fact, the scale of the continents that overtakes nations and businesses.

Although invisible, the cloud is not without place; rather, it is the instrument through which the places acquire (through the deployment of the code) their ontology and manage a new sovereignty. The cloud infrastructure serves and is served by logistic networks of different nature that move physical objects as much as data and metadata. In doing its job, the cloud constantly absorbs data through which it simultaneously centralizes and decentralizes people, sensors, services and devices (recently, globally, the total amount of data stored in the cloud has exceeded that stored locally).

The cloud layer governs and exercises its sovereignty over time, space and subjectivity, the central dimensions we have analyzed about the new way of experiencing. To give a concrete example, it is the cloud layer that synchronizes our devices and the content that we use on different devices such as PCs and cellphones, which configures authorized access to offices, factories and laboratories, which manages our identifying credentials within the platforms.

The city

This is the layer that includes the environments and the emerging relationships of the metropolis and the so-called megacities. It identifies urban space as a social and spatial territory that is algorithmically constructed to enable human endowment, but also its mobility through a mega-network of logistics infrastructures. Cities are increasingly becoming independent entities, with autonomy and intelligence, as can be evidenced from the emerging smart city paradigm. It is worth recalling that 80 percent of the countries' gross domestic product is produced, in fact, in metropolitan areas.

Megacities have a morphology that intersects the highest layer of the cloud and its platforms with the intermediate and closer interfaces and users. Of this codified city, a grid of roads and blocks is what we see and understand because we still use pre-algorithmic and pre-protocol concepts and models. But now, like Neo after he swallowed the red pill, we should be able to look at urban spaces as reconfigurable instances of code that now accelerates and now immobilizes physical objects and data packets, even as it negotiates the boundaries between personal and public areas.

In fact, the design of this urban interface feeds on data in real time and is dynamically structured through the use of code both for mobility and for permanent conditions, both for private and personal spaces and for those with an extended social network. It is a programmable city with programmable environments and buildings; as for roads and mobile networks, buildings also take advantage of sensors and intelligence. We will evolve from living *in* a house to living *with* a house – say Ratti and Claudel – with hardware/software architectures that will take on the ontology of the interface and which we will use to interact;[5] vice versa, other structures of the classical urban landscape, such as traffic lights or car parks, are destined to disappear with the introduction and adoption of autonomous vehicles. In a stacked world, the algorithmic urbanism of immobility and mobility orchestrates the whole city as a single invisible interface.

The address

This is the layer that identifies objects, interactions and transactions in a univocal manner through universal, massive and granular addressing systems such as the IPv6 protocol.[6] This layer makes sure that every object or event appears on the stack as an entity with which to communicate and, therefore, potentially interact with a computable and logistical interface. At different spatial and temporal scales, objects, places and subjects are also in possession of a static or dynamic network address. The progressive adoption of the IPv6 protocol means that, if necessary, each object can have its own unique address. The geopolitical sovereignty of the pile is not limited to the typical cartographic division of the planet, but is completed with the ability to construct a synthetic and artificial cartography of identities localized for things, people and events. Thanks to the addressability based on the identification of individual entities, hierarchical bifurcation systems and tables of address resolution, the stack knows of the existence of everything within its domain (the recent cyber-attack mentioned in the introduction has affected this aspect).

And everything "exists" in this new ontology, as it has a unique address within the grids of the stack. In this scenario, the computation ceases to be the specific equipment of a machine and becomes, within a sociotechnical environment – a milieu – a property of the things of the world. In this way, all the objects become computational machines. We still call them cars, airplanes, televisions, toasters, refrigerators, homes, but in reality, they are and will all be "peripherals" of this accidental megastructure. Also, they become sensors and nodes of a sensor network. The logistics of constant contact is based, in full, on the ability of the stack to know, in every instant and in real time, the state of things addressed. And with the industrial internet, not just of the things produced, but of the production itself. And, again, with blockchain, not only of things, but also of every value, asset and good on the internet.

The interface

This is the layer that includes the modes and devices with which users mediate and are mediated inside the stack. Such interfaces may be of a different nature (kinetic, tactile, gestural, semiotic, with AR or VR), to the point where even the interface may not be visible to the user as such. The content of the interfaces can be considered a moment of access to the information infrastructure of the stack and its artificial and collective intelligence. Interfaces are the tools with which users perceive the world (say, the immediate display of an image on a mobile device). At the same time, they give the opportunity to users (humans and non-humans) to be recognized by the stack and to feed the different top layers that we have described with data and metadata.

This exchange takes place according to multiple lines of interaction, even if occasionally contradictory. It can be promiscuous or prophylactic, physical or digital, accelerated or immobilized, territorializing or delo-calized. And we are witnessing an evolution of this interfacing that is progressively being freed of the most classical peripherals (the screen, the keyboard ...) to invent interfaces directly constituted by the objects and the programmable and intelligent materiality of which they are made. Not only does the interface multiply and make itself invisible, but it becomes increasingly intelligent and conversational.[7] The application of artificial intelligence is rapidly changing the concept of interface to which we have been accustomed in recent years. As we have said, interfaces work and will work more and more in a feed-forward perspective, in anticipation of time (and therefore of intentionality and behavior). And more than designing them to speak to us, we will design them to speak for us as our agents and artificial assistants. But who, exactly, is *us*?

The user

That is the layer that represents the subjects acting inside thanks to the stack. Relevant here is how the stack sees users (human and non-human, individual and collective, dividual and infovidual), through which are identified processes of quantification (increasingly through digital and artificial technologies), of qualification (platforms, protocols), behaviors and subjects. The user as a layer of the stack is not to be equated with the concepts of individual or *person* (which, today, still represents the key subject of research and design thinking practices), but a model not given in advance, built through and by virtue of the interfaces and platforms that intersect the various other layers of the stack.

More than subjects in the classical sense, they are instances of instant agency, situated and contextual, in many cases fragile and emerging technologically. Subjectivities that are built, but also continuously overwritten and updated. They are infovidualities – as we have learned. And when an intentionality collapses into an actual interaction (such as a click for a purchase, a posted image, a telephone call, an entry into a sensorized building, a quantified heartbeat, and so on), the profile of that subjectivity is modified, also transforming future potential actions. As, for example, when we use our Facebook profile to sign up for another platform like Airbnb.

At the same time our profiles and the probable profiles of other users who are related to this event are also updated. To say that we are online only when we imagine accessing the network is an obsolete way of considering individuality in a digital and artificial world. Even if we are not online, in reality, our data and algorithms constantly feed and modify our own and others' profiles, preferences and probabilities. The information philosopher Luciano Floridi introduced the *onlife* neologism[8] to signify that the online/offline dichotomy is now outdated.

And platforms are built precisely on the economy of the administration of these *infovidualities* through data and algorithms. The concept of a platform is that which keeps the different layers of the stack intersected. At this point, the matter requires a specific investigation. But what is a "platform"?

Platforms: hybrid species

A growing series of analyses are asking today what is a platform, considered not only and not so much as a technological reality, but above all as a new organizational and business model.[9] Here too we try to deal with the phenomenon through the lens of philosophical speculation.

The most abstract definition of what constitutes a platform is the Bratton proposal where, in his chapter *"Platform and Stack: Model and Machine,"* he writes:

> Platforms are what platforms do. They pull things together into temporary higher-order aggregations and, in principle, add value both to what is brought into the platform and to the platform itself. They can be a physical technical apparatus or an alphanumeric system; they can be software or hardware, or various combinations. As of now, there are some organizational and technical theories of platforms available, but considering the ubiquity of platforms and their power in our lives, they are not nearly robust enough. Perhaps one reason for the lack of sufficient theories about them is that platforms are simultaneously organizational forms that are highly technical, and technical forms that provide extraordinary organizational complexity to emerge, and so as hybrids they are not well suited to conventional research programs. As organizations, they can also take on a powerful institutional role, solidifying economies and cultures in their image over time.[10]

To all intents and purposes, we will be faced with a new hybrid species: the one that, coining a neologism, I have defined *platfirm*,[11] crasis for the note of the two terms, *platform* and *firm* (company).

The platform can be based on an open global distribution of interfaces and users (and in this it resembles markets); at the same time, however, the coordinated programming of interactions between users and participants on the platform makes it similar to hierarchical organizations. This would be a techno-economic system based on a standard that simultaneously distributes interfaces through their remote coordination and centralizes their integrated control through the same coordination – Bratton continues. All this is managed by huge amounts of data. The data constitute the foundation of this new platform capitalism.[12]

To have an idea of how this phenomenon has burst on to the scene, we only need to remember that the first five companies by market capitalization currently are, in effect, platforms: Apple, Alphabet, Microsoft, Amazon and Facebook. During the last fifteen years – as we know – historical companies such as General Electric, Exxon, Citigroup and Walmart have been removed from the ranking of companies that embody, with specific dynamics, the logic of the platform: companies strongly technological and software-driven, able to scale rapidly and break the boundaries between industrial sectors and business models. To give a recent example, after the request and long-term attempts to become a global financial institution, Facebook has obtained from the Central Bank of Ireland the authorization to operate as a payment service company in European countries (in the USA it was already possible via Messenger). It will become an electronic money-issuing company offering, for example, the possibility for members to transfer money abroad and, along the way, many other connected services. The *platform disruption* is so strong that many today also wonder, among other things, if the classic divisions between industrial sectors remain valid.

The platforms are mechanisms, engines or if you prefer generative software programs that draw the forms, dynamics and rules of participation, in accordance with technical, discursive and performative protocols. In a business perspective, I like to define them as a "new organizational and business model of plug and play that allows a series of beneficiaries on the Internet to co-create and exchange value." To give an example, Airbnb as a platform brings together people who make a room available to other people who are looking for accommodation; while Uber crosses the offer of rides with the demand for mobility on request. The interaction via the platform co-creates the value that is exchanged between the different participating beneficiaries.

The philosophical lens, of course, looks at these dynamics in deeper ways.

If we try to summarize Bratton's most abstract principles of an emerging *platform theory*, we can identify them as follows: *a)* the platforms design and allow interactions between the participants, but remain open to unexpected behavioural occurrences; *b)* provide for a rigorous standardization of scale, duration and morphology of their components; *c)* the standardization produces a consolidation of the system and a disincentive for users to invest their time and energy in other platforms; *d)* standardization can be reprogrammed by users to allow innovation, even in terms of guardrails and restrictions; *e)* the design and governance of the platform are based on formal models for predicting and organizing information; *f)* the value generated by the user's input/interaction, both for the user and for the platform, must be greater than the mediation costs incurred; *g*) the platform must manage the balance between centralization and decentralization; *h*) identities that allow access to services are generated for users, both human and non-human; *i*) the platform is able to manage identities in an extremely individual way by assigning domains and even temporary accesses; *l*) the platform is an empty information diagram or with minimal information waiting to receive and manage the requests coming from the users; *m*) any structural component of the platform, such as protocols or interfaces, can be replaced; *n*) the platforms govern both instantaneously responding in real time to the interactions and cumulatively in relation to certain analytical and profiling thresholds; *o*) the platforms function as distributed sensor systems that encourage the identification of errors or anomalies, thus interpreted with respect to the formally configured models; *p*) platforms that reorganize existing systems and information tend to achieve generative rooting faster; *q*) the ubiquity of the platforms makes them more robust, but also more vulnerable and prone to destabilization; *r*) the platforms do not resemble the work they do and do not do the work for which they seem to be configured; *s*) the sovereignty designed and achieved by the platforms is in many circumstances automated, but in others it can also be contingent and strongly susceptible to interference.

From a business innovation perspective, a platform is a new "operating system" at the crossroads of free markets and organized hierarchies able to connect (with rapid mode via API, for example), to interact (in synchronous and asynchronous, human or algorithmic mode) and co-create and exchange value (the value is produced in the platform, not by the company) among multiple network beneficiaries (non-users or even simply producers/consumers), using resources and assets that can be proprietary or non-proprietary, physical as well as digital. To give a sense and an example of the difference from usual business strategies, we could say that platforms tend to scale on the demand side (using network effects), while traditional enterprises are used to scale on the supply side.

In doing this, a platform (a *platfirm*) is: *i*) an engine of continuous innovation for business models; *ii*) an architecture that exponentially enhances assets, goods and resources external to the company; *iii*) a reconfiguration of what is value, the types of market and user behaviour; *iv*) a low-friction system in the distributed and constant management of transactions and interactions; *v*) a game changer for traditional rules and practices of leadership and management.

These platforms, which in the third industrial revolution were designed to be entirely digital, with the fourth industrial revolution are assuming the dimensions and dynamics of atmospheric and pervasive architectures, intimately connected to the physical world (think, for example, also to the recent developments of Airbnb that, from a digital marketplace that brings together supply and demand for rooms, is now proposed as a creator of real travel experiences). And these new services do not always support conciliatory positions among them. Platforms seek market and social dominance within the stack. For this, they are willing to trigger heated confrontations, even, in real competitive conflicts, overriding legal or strategic, evident or more subtle provisions.

This is the stack in its current configuration. The one to come, to the construction of which we are all called, will be the fruit of negotiation that we will be able to constantly exercise, as citizens and consumers, with respect to the paradigms and technological vectors that are in action and those arriving. Such as blockchain ...

Network, stack and chain

In recent decades, we have become familiar with the metaphor of the network and, in this chapter, we have begun to look at the world through the speculative lens of the stack and its platforms. But there is another nascent construction of contemporary thought that is beginning to appear as protocols, architectures and algorithms. For many observers, it could revolutionize culture, economy and society with the same (and perhaps even greater) impact as that of the web. We have already encountered it an our journey. It is the blockchain.

It is occasionally compared with the invention of the container in the late 1950s: before that invention, the logistics and the transport of goods were expensive, labor-intensive, non-standardized and time-consuming. With the introduction of the new system of transport and exchange of goods, in the following twenty years, global trade would expand by several orders of magnitude. The blockchain would represent, within a new distributed ecosystem, a new logistics of value and goods able to grow exponentially a new economy.

We have given it a succinct definition, but now we need to complete the description and the analysis because its arrival and its potential adoption could modify, again, some of the concepts that we normally use such as trust, identity, truth, event, proof, interaction, etc. So well beyond just moving goods digitally, we would be faced with a new institutional procedure (in the sense of a set of formal and stringent dynamics) able to reduce the uncertainty ratios in markets and organizations. Not just a technology or an infrastructure, but a new social institution (after the market, the enterprise, the nation).

As Mougayar, author *of The Business Blockchain*, wrote:

> Understanding blockchains is tricky. You need to understand their message before you can appreciate their potential. In addition to their technological capabilities, blockchains carry with them philosophical, cultural, and ideological underpinnings that must also be understood. Unless you're a software developer, blockchains are not a product that you just turn on, and use. Blockchains will enable other products that you will use, while you may not know there is a blockchain behind them, just as you do not know the complexities behind what you are currently accessing on the Web.[15]

Historically, the blockchain is the second major technological layer (after the arrival of the web in the mid-1990s) with which we can build digital solutions and services on the Internet. As the web has had its creator in Tim Berners-Lee, blockchain too has its own founding genius. He is called Satoshi Nakamoto but he doesn't exist. Or rather, his/her (their?) true identity is shrouded in mystery. Under this pseudonym, a paper presenting blockchain as the root of the innovation introduced by the Bitcoin virtual currency was published in 2008. But it is not just a question of virtual currency or crypto-money, as it's called. We cannot go into the technical details here. In reality, the deeper technological aspects of the new protocol are not necessary to understand the conceptual principles and potentialities of the new architecture, and, to follow later, many digital services of a radically new nature. Technically,[16] it is a network that maintains a public and encrypted distribution ledger; in terms of business it is an exchange network to move values and assets, not only information as has so far been the case for the web. From a legal point of view, it represents a mechanism and a dynamic to validate transactions without resorting to the intervention of trusted third parties: a notarial document, a marriage contract, a monetary exchange, an identity certification, a financial transaction, the purchase of a car are all acts that, so far, have been registered on databases (once paper books) that a centralized authority (the banks, the religious authorities, the governments, the service companies and so on) was called upon to validate.

With blockchain, all these trust entities would be disintermediated thanks to a shared, encrypted and distributed digital ledger that aims to preserve publicly various kinds of activities, to protect them against attacks or falsifications and to optimize interactions and transactions. This would be the second wave of disintermediation coming after the first one introduced by the web.

It should be clear from this first description that it is not yet another technological protocol that operates on the Internet. Employing software engineering, cryptographic science and game theory, the Internet may have found what has been missing so far: a protocol to manage *trust* in relation to the exchange of values and not just information. This is why blockchain has also been defined as the *trust machine* or the *trust protocol*.

How does blockchain create this trust and what is its ontogenesis? In the blockchain, trust (to the veracity of any transaction, event, or action) is created through a decentralized consensus. This consent transfers authority and truthfulness to a network of nodes that are committed to register and encrypt the transaction in a chain of blocks of information that is publicly verifiable and algorithmically executable. In this way, the blockchain subverts the theory and the openness with which the concept of trust is designed and commonly defined.

In fact, the current intermediation of the centralized entities, which by tradition preserves, certifies and gives confidence, presents critical and inefficient economic and social factors that the new decentralized, encrypted, algorithmic way could eliminate. If we think about it, when we access our bank account, we actually access a database that identifies the source and recipient by changing one state with another.

Centralized trust is expensive (think of bank costs, credit cards, notary fees...), slow (think of the time it takes to transfer money or financial transactions), attackable (many malicious data intrusions plague government data centers and private institutions), opaque (transparency in operations and policies raises much criticism) and discretionary (a wide margin of discretion is today permitted in bureaucratic and centralized institutions).

Blockchain is called upon to take on these inefficiencies – some of which are intended, others unintended – by resolving them with an open, distributed, coded and encrypted approach. This new protocol can be used in any context where it is necessary to allow a certified, safe, accelerated and inexpensive exchange of value. If we can imagine for example that, with the IoT, there will be billions of objects that will sense the world and act in the world communicating data between them and with us – to buy energy on our behalf, to protect our health or to save us time and money – blockchain technology would seem to be a good strategy for privacy protection and for data sharing with security and efficiency.

Finally, speaking of sovereignty and governing, blockchain is also suggested to build new organizational and governmental forms. Some people imagine using blockchain to even disrupt the kind of organizational models we have known them until now. These extreme theorizations are called DAO (*decentralized autonomous organization*): in substance and in short, from voting procedures to making collective political or social decisions, the idea is to use smart contracts to make humans and machines interact in an automated, decentralized, protected, rapid and inexpensive way, surpassing the traditional hierarchical organizational dynamics (slow, inefficient, corrupt, opaque). Raval[17] speaks explicitly of decentralized applications stressing the concept of node parity (no node instructs the others), while the standard concept of distribution does not imply this kind of parity (in fact, in many cases, the distribution of the computation remains centralized).

Contractual relations algorithmically predetermined, encrypted, decentralized and self-executable would replace historical, private and public organizations, regulated by traditional communication and negotiation mechanisms. In the future, the idea is to make human organizations evolve toward more participative forms (voluntary affiliation), collaborative (oriented toward common objectives), cooperative (including sharing the generated value), distributed (thanks to the propagation and multiplication of nodes of extended networks), decentralized (in highly scalable mode) and automated (thanks to algorithms and artificial intelligence for self-sustaining and to share value in an equal way).

In addition to the financial, health and energy networks, human organizational networks could potentially be affected by the new protocol and its embodiments such as smart contracts. In these scenarios, blockchain technology would seem to be the ideal candidate for building a safer, more comfortable and fairer future. I wrote "candidate" not by chance. Thanks to the work of several startups, the potential of the blockchain protocol has begun to be understood and financed. Banks, financial institutions

(MasterCard and Visa began investing in the blockchain at the end of 2016) and governments are now beginning to experiment with the new technology, because they perceive the benefits in terms of service performance and cost reduction, leaving for now at the margin the potential for disintermediation.[18] At the same time, as for all strongly disruptive technologies, the potential is at least equal to the risks. And this applies to many of the disruptive innovations and perspectives we have discovered in our journey. Recently landed in this list of risky anxieties, is artificial intelligence. But are we sure that talking about risks is the right approach? Given the expected potential, are we sure that fears, resistance or waste are the only alternatives to protect the human species and the entire planet?

Vulnerability beyond risk

Recent blockchain attacks (as well as the cyber-attacks mentioned at the beginning or the software failures in self-driving cars or the theft and misuse of personal data) are emblematic of the kinds of objections raised to date regarding more generally the complex times in which we are living. These vary from unstable innovation linked to the nature of the software, or the fallibility of the code and its invisibility, or again the use of algorithms based on collective and unpredictable artificial intelligence, to the danger of total archival and privacy of citizens and consumers. Our time is open to extraordinary possibilities of inventiveness, resolving old problems and providing new opportunities for economic, social and political improvement. These are times, however, that are increasingly described in terms of risks, such as those for employment taken over by intelligent machines – so they say – but more generally, because of the deployment of code, data, algorithms and intelligent agents that can jeopardize the human species. Also on this point, however, I would like to propose and launch a more informed perspective on the risks of innovation.

I believe that, aside from the concept of risk, we should try to reason, more constructively, in terms of "vulnerability". In fact, risk is the paradigm and framework that today drives the reflections around the technological revolution we are going through. The risk of acceleration, the risk of the *singularity*, the risk of robotics, the risk of artificial intelligence, the risk of privacy , the risk of cryptocurrency and so on. Fear of risk, risk analysis, risk management are the dimensions that often meet and clash in the public debate as well as in corporate meetings. Risk is the dominant paradigm. Risk, however, cannot exhaust the complexity of the new situations in which we will live and should not be the only approach. More often than not, risk is confused with other aspects: contexts of ambiguity (the risk is calculable, but without agreement on the consequences) or of uncertainty (there is convergence on the consequences, but the impact cannot be measured) or ignorance (there is no homogeneity of views nor is the risk quantifiable). For the sake of clarity, risk should be reserved, then, only to situations in which there is unanimity of judgment between the parties and a precise measure can be given of the consequences of a given action. In the past twenty years, however, the risk paradigm has been put into question precisely because of the very uncertainty deriving from a series of disruptive technological innovations. Think about genetics and the manipulative experimentation of the genome, human and non-human, about the introduction of nanotechnology with its interventions on the micro and subatomic matter, about the massive introduction of the artificial algorithmic and robotic intelligence or, more recently, about the arrival of quantum computing.[19] In all these cases, the lack of long-term prior experiences leaves the risk experts a bit in the dark. Laboratory experiments cannot reassure completely as to the scalability in space and time of the results of tests and experiments in vitro. Nor does a single assessment or risk management by technicians and scientists suffice, as it very often concerns normative subjects upon which, among other things, different cultural models can be applied. And the same difficulties involve not only the risk analysis but also its management.

In the face of these problems, a new horizon of interpretation appears better able to deal with the complexity of the programmable world that we tried to describe: the paradigm of vulnerability.[20] The new paradigm has emerged from many domains: economic analyses on poverty and life perspectives, climate studies, research on the criticality of technological infrastructures, investigations on environmental disasters, anthropology and theories of development.

But what is vulnerability? Each of the domains mentioned has, of course, its own definition that goes from the human being's capacity to face stress and change to the probability of occurrence of events impacting on a specific context. Some definitions focus on the ability to respond to stress conditions, others on the conditions that produce vulnerability such as the occurrence of the unexpected.

If this is the horizon of meaning, vulnerability would seem to be the condition of our time. Humans are vulnerable as are plants, animals, ecosystems, groups and communities, human institutions and artefacts, technical systems and organizations. Vulnerability is an emerging property of complex systems and as such enters the analysis of our socio-technological condition. It is therefore a question of moving from the consideration of a risk society to facing the dimensions of a vulnerable society. How is this done?

The first indication is that the concept of vulnerability is not purely negative. A certain degree of vulnerability is accompanied by openness in which some level of uncertainty, linked to learning and innovating, is necessary and inevitable. Vulnerability is a broader concept of risk and can include situations where outcomes and contexts are more difficult to quantify. The concept of risk has a decidedly more quantitative approach. Risk estimation, as a rule, is the probability of hazard multiplied by the impact of its occurrence, and so is usually treated by engineers and nature scientists. But we also need a more qualitative and open interpretive horizon that requires conceptualization and critical reflection. Risk and vulnerability also employ two distinct vocabularies: that of risk is clearer and better recognized, that of vulnerability more nuanced and contextual, hence less precisely defined.

A distinct vocabulary gives a better sense of this difference between risk and vulnerability: to "prevention" is opposed "precaution," "indeterminacy" to "surprise," "procedure" to "prudence," "uncertainty" to "unpredictability," "security" to "solidarity," "institution" to "community" and so on.

Take the case of personal data and their protection. In fact, many people believe that, to obviate a breach of privacy, they need to prevent or limit the collection of information and data. In fact, as many jurists point out in discussions on this issue, the protection of privacy will more likely require, counter intuitively, the collection of more data and not less data, as well as redesigning a concept of privacy that may be more aligned to the new ontological contexts produced by the digital and artificial technologies we have talked about so far. On the other hand, the opportunity, for example, to use the massive data on health and behavior of citizens to advance medicine or improve public health policies are very high and should be pursued. How should opportunities and threats be balanced?

A good example of an approach that tries to balance emerging opportunities for technological innovation and attention to vulnerability is what Alex Pentland and his research group are promoting with regards to personal data management. Acknowledging that personal data have become a real individual and social asset, researchers are elaborating and testing a new framework called the internet of trusted data.[21] The model consists of four fundamental strategic and operational principles: *a*) a robust digital identity, that is to say, the creation of certified, secure digital entities, *personas* that are associated with different personal sets of validated data (my digital working self, my health data, my citizenry and so on) to which each of us can access and that each of us can verify and decide to share with third parties; *b*) a distributed network of trusted authorities able to implement a degree of secure consensus on data and decisions, perhaps using the logic of the distributed and secure register of blockchain; *c*) a distributed system of computation and protected computation in which the algorithms – made available in open and readable form – go to data (and not vice versa, thus reducing the risk of attacks) that must always be preserved in encrypted form; *d*) the possibility of universal access, controlled by citizens and security policies, to personal data accessible through interfaces and devices that facilitate their use for the benefit of individuals and the community.

This is a concrete proposal that is perfectly in line with an approach to vulnerable societies and which seeks to create the most open conditions possible for the maximum enhancement of technological innovation (in this case, the increasing availability of personal data that individuals and society can benefit from, but safeguarded by provisions of control, protection and attention procedures so as to avoid potential threats deriving from the massive availability of private information on individuals, families, human and social groups. As mentioned earlier, paradoxically, we will need more data to protect our people and lives and not less data (and avoid rejection and closure against the potential risk).

To govern a vulnerable society – which has the ontology we have reconstructed throughout this text and which we will summarize in the concluding matter – is not and will no longer be a simple operation. We will certainly need to develop as best we can an ethics of vulnerability and an ethics of data.

Now, on the other hand, we can also see the why and wherefore. We have gradually unraveled many of the knots that prevented us from understanding the whole. To conclude, let's try to recall these steps briefly: what we have here is a software code that, due to its constitution, is in a state of continuous deconstruction and reconstruction of itself with outcomes that are not entirely predeterminable, a new apparatus of sensorial networks that engages in the environment and co-emerges creating unexpected techno-ecologies of experience, the invention and creation of alien intelligences that, by algorithmic means, understand and act the world in ways that are no longer totally intelligible to us, a production of data that redefines the ways in which we feel and we think about the world and that becomes the ultimate interface through which, in an unconscious and anticipated way, we inhabit a programmable reality and, finally, an ecosystem of all these dimensions within a stratified politically oriented architecture of governance that recognizes and modulates (through platforms) resources, processes, and social and economic dynamics.

The emotional and cognitive commitment of everyone, including philosophers, will be necessary to ensure that this new phase of the Anthropocene, as the human era is today defined on the planet Earth, can produce a truly better programmable world for all.

Notes

[1] "Biopolitics" is a term used and popularized by the philosopher Michel Foucault to indicate practices of power that result from direct implication on the part of politics and that of life understood in its strictly biological characterization (for example, the body).

[2] Yuill (2008), p. 165.

[3] Illuminating the point, see Mackenzie (2006).

[4] Zero UI indicates the approach used to design interfaces that do not appear so, being so coextensive with objects and environments. Our gestures, movements, sound emissions, eye beats with saccades and fixations (rapid eye movements and pauses between them) interact directly with objects and environment without intermediation. Screens to be seen and buttons to click – classic interfaces will disappear – to give way to an intelligent environment that feels and responds to our natural presence and behavior.

[5] Ratti and Claudel (2016).

[6] The infrastructure that makes it possible to uniquely identify the nodes (devices, computers, accessories) of the network was based on the IPv4 protocol, a model however put in crisis by the continuous growth of the number of objects connected to the network. From 2011 onwards, a technological change is taking place that is progressively leading to the adoption of a new standard, IPv6. Compared to its predecessor, IPv6 allows an infinitely greater number of devices to be connected to the network (for example, the IoT or the industrial Internet and the need to have unique network addresses for billions of objects).

[7] An excellent summary of the state of the art of conversational interfaces can be found in McTear, Callejas and Griol (2016).

[8] Floridi (2014).

[9] The phenomenon of the emergence of a platform-based economic model is now widely reflected in the publication of numerous essays, to the point that we have come to speak, more generally, of Platform Capitalism, as stated in the title of Srnicek (2016). Among the many volumes released we mention: Choudary (2015); Parker, Alstyne and Choudary (2016); Evans and Schmalensee (2016); Moazed and Johnson (2016); Claire and Reiller (2016); Shaughnessy (2016); Libert, Beck and Wind (2016); Roth (2016). Simplifying, the analysis of a platform economy has, today, a dominant orientation and another we could call alternative. The first (platform capitalism) supports and encourages the use of platforms, favouring the ability to organize underused resources on a scale, crossing the demand and supply markets as best as possible. The second (platform cooperativism) believes that the new distributed architectures should, instead, privilege the sharing of goods and property and have a more democratic governance structure, on a par with the actors of the ecosystem (today unbalanced in favour of platform owners). Technological neutrality of access to services is not sufficient. The forms and dynamics of ownership and participatory governance of the platform are also relevant.

[10] Bratton (2015), p. 41.

[11] Working on the preparatory materials for the eighth edition of the 2015 Social Business Forum organized by OpenKnowledge in April, in a burst of inventiveness, I created neologisms trying to best express the organizational and business innovation that was emerging in those months. Among these, I had imagined the term *platfirm*. These new words formed, in July 2015, the key content of my keynote at the Forum.

[12] Srnicek (2016).

[13] If the industrial revolution had the factory as its propulsive engine, many today believe that the "platform" is the equivalent for the post-industrial one – a new era of platforms represented by the success of native platform-companies like Amazon, Apple, Google, Airbnb, Uber, but increasingly also by initiatives of traditional companies such as, for example, Philips (with its HealthSuite connected health platform) or Disney (with the MyMagic + consumer engagement platform) or General Electric (with the cloud platform for the Predix industrial environment) that do business by adopting, in fact, the models and practices of a platform. In this sense therefore, talking about platforms is not referring to architectures, technologies and applications, but to new strategies with which companies can organize themselves and to new models with which managers and executives can do business and compete. The platform provides an open, participatory infrastructure for these rule-governed interactions. The purpose is to facilitate the match between users for the exchange of goods, services or social currency, enabling the creation of value among all participants. The reasons why the platforms disrupt traditional companies reside in the fact that they subvert the classic organizational and business forms, exporting functions, processes and resources. Use of network effects, high scalability of operations, use of non-proprietary assets, reintermediation and aggregation of markets, community involvement are just some of the success factors of this outsourcing. The *platform design*, as described by Choudary (2015), identifies how to make a platform attractive (*pull*) for its participants, how to simplify and reiterate the participation (*facilitate*) of users within the platform and how to optimize the exchange of value between producers and consumers (*match*). Key concepts are: the focus on interactions that you want to enable (*interaction first*), the simplification of access and participation in the platform (*frictionless entry*) and the application of filters to increase the importance of co-creating value among the members (*algorithmic filter*).

14 Bratton (2015).

15 Mougayar (2016), p. 1.

16 In order to provide a simple and concise description of the technological functioning of blockchain we can say that it is a network that functions as a mechanism to validate, store, update and protect transitions of states of reality (transactions, interactions, events and so on). In the blockchain register (*ledger*), every event (for example a payment) has a *timestamp* (temporal indication) that irreversibly tracks the state transitions that occur (a database can be modified, a register remains unchanged). Each event is encrypted in an information block and added to the list of previous events (*blocks*), forming the block chain (*blockchain*). To be inserted in the blocks and then into the chain, an event (for example a transaction) must be validated. The validation is carried out by special network nodes called miners that, thanks to considerable computational capacities, solve complex cryptographic operations in order to find the solution (proof of work). The first node among the miners that finds it is rewarded (in virtual currency) and enters in the blocks of the chain the new state transition (the event to be recorded in the ledger in an irreversible way) that contains traces of the code of the previous block (*hash*). Finally, this updated log is shared among the network nodes that keep a copy of it, limiting the risks of cyberattack.

17 Raval (2016).

18 It is no coincidence that many of these developments, sponsored by banks and financial institutions, seek to build private and editable blockchains – infringing in this way two of the founding principles of the new protocol: its openness and its irreversibility. The story seems to repeat itself here with new protagonists: as it happened with the Internet at its beginning, when the telcos tried to build a private internet.

19 To give an example of the complexity and vulnerability associated with the introduction of quantum computing, we recall that many of the cryptographic keys (and also, to be clear, blockchain and with it virtual or crypto coins like Bitcoin) that today are used widely and with confidence to protect data and information of private companies and public institutions would be put hard to the test and eventually attacked by machines now able to calculate with exponential powers. In this perspective, it is clear that we need to work quickly to create cryptographic keys that are appropriate to an era defined as post-quantum computing.

20 The work edited by Hommels, Mesman and Bijker (2014) documents this paradigm shift through a series of theoretical contributions, as well as operational cases, both private and governmental. This paragraph synthetizes their book perspective.

21 On these issues see Hardjono, Shrier and Pentland (2016).

Conclusion

Toward a Proto-Data Age

A strange world, one that is simultaneously artificial and natural.
Y. Hiu, *On the Existence of Digital Objects*

Looking back now on the journey traveled so far, I think we can say that there are three key dimensions with which to confront our digital and artificial future: *programmability, invisibility* and *sovereignty.* Sensors, data, algorithms, artificial intelligence and platforms share all three dimensions, albeit in different degrees and scales. By programming our new reality sovereignly and invisibly, these technologies are redefining in depth concepts and philosophical perspectives (from anthropology to ontology, from ethics to epistemology) and with them our understanding of the world.

It was a rapid and – I guess – an intense exploration aiming to find the answer to the initial question about the ontogenesis of our present and future. What, then – to return to the provocation of Jarzombek – of the ontology that is emerging because of the writing of the world by the code, data and algos? What are the salient features of this sentient, quantified, algorithmic, artificial and synthetic new civilization?

Seven surprising A's

I have tried to summarize the seven dominant orientations that we can identify as intersecting ontogenetic vectors, in a systemic way and reciprocal reinforcement: *amplification, automation, adjournment, anticipation, alienation, anarchiviation, atmosphericity.* I imagine a new cultural and philosophical investigation that can, in the future, offers its contribution by entering into the heart of these dimensions.

Not as one, but as many

The first of the ontogenetic forces is linked to the capacity of new technologies and new systems to promote extreme amplifications. The frictionless reproducibility typical of digital technologies has expanded further to embrace multiple dimensions and not just information. It is the ontology of the exponential multiplication of temporality, of spatiality, of subjectivity, of experiences, of intelligences, of data, of ecologies, of sensoriality, of software code. In this perspective, artificial intelligence is the latest *inflationary (media) technology* to arrive. Having the ability to morph the hidden path of the media and the world (and not just to duplicate a copy as for the digital technology), it represents a new, qualitatively different, level of reality (with its own, new truth/false regimes to be learned).

Toward a squared automation

The second generative dimension is based on the digital and AI technology networks and the potentials that rely on automation. We live in an *automatic society* as Stiegler define our contemporaneity. It is a dimension that involves many different businesses and markets: from marketing automation to autonomous vehicles, from automatic contracts to algorithmic finance. There is no possibility of amplification without the ability to engineer automation going as far as squaring automation with technologies such as machine learning. And lately, this automation will impact even the code production itself. With *deep learning*, in fact, no human is involved in writing the code. It's the code itself that, using data, searches the space solutions for a program that satisfies the constraints.

To be is to be updated

The third major vector supported by technological architectures and operating models with continuous design and delivery approach is the update. Being updated (for a system, an application, a platform) is a vital condition for a given entity to be in the world. Forgetting to update, vice versa, puts at risk the existence of things in the world and many related concepts (eg. identity, agency, accountability). This model has also many different ontological implications. Think about the concept of technical failure. We are used to consider failures as momentary events. But because of continuous updating and software fallibility, we have to rethink failure as unfolding, as a continual repair process. We need to design for failure and to operate in order to learn. Failures will continuously happen, so we must increase our ability to rapidly repair faults diminishing the cost of failure at any level. The update cannot be achieved without automation and amplification.

The age of precognition

The fourth ontogenetic condition that emerges from the intersection of data and algorithms is anticipation: the ability of the new sensorial and cognitive apparatus to anticipate (with feed-forward mechanisms) events and behaviors. We are not in the age of cognitive computation – as you hear it called – it is the *age of precognition*.

As humans, in fact, we need to manage not only the present information overload, but the future world uncertainty. The prolepsis, the orientation toward the future, requires amplification, automation and constant updating (even if invisible to the human as they are produced by autonomous and automated technologies). I often say that "the production of prediction is the very business of infonomics." To name a few: media platforms sell predictions about people's behavior, transportation industry has transformed traffic management in a prediction business, precision medicine is starting to work predictively for curing diseases.

Think like the aliens

The fifth ontogenetic dimension is connected to the explosion of the technologies, techniques and algorithms of artificial intelligences. More generally, the new world is emerging thanks to the application of new types of "intelligence", able to act autonomously and creatively leaving the human and its sensory and cognitive abilities out of the loop. AlphaGo invented a new, unexpected move to win against Lee Sedol breaking this human local maximum. Many industrial products are now designed using this sort of algorithmic imagination. An imagination computationally able to mine the infinite space of possible solutions and to find creative and non-human answers. This alienation (understood as the transfer to others agents of possibilities and rights) takes on different dynamics: infovidualities, artificial lives, smart sensors networks will negotiate intelligence and autonomy.

From archive to anarchive

The sixth guideline that characterizes digital and artificial worlds is a new kind of archive (and archiving methods) that, despite having the function of recording, deviates from traditional archiving operations. We speak of *anarchive*. In our perspective, there are two new elements of differentiation: the first concerns the trend of "total archive" (not partial as for the old archive concept), the second refers to the trend of the "dynamic archive", oriented to make things happen (that is, to activate our world, not just to preserve it). The blockchain protocol is the latest example of anarchive (both, total and protentive) able to record past transactions as well as to execute future contracts. To date, blockchains use cryptographic technologies to archive the totality of socio-economic interactions in a decentralized ledger and use programming languages to run self-enforceable legal contracts in form of scripting code.

Atmospheric mediations

The seventh basic element is the emergence of atmospheric media that, by short-circuiting scales along with micro and macro dimensions of sensitivity, build new experiences and a new manner of subjectivity. Think about gestures interfaces, sounds and voices, waves or phases for smart speaker interactions. Think also about aerial drones, virtual and augmented realities or holographic visualizations. If this is true, we have to move rapidly from outdated concepts such as media contents, media channels, media users to new and surprising concepts such as *phase perturbations* (Ash), *feed-forward sensibilities* (Hansen) and *elemental media* (Durham-Peters).

In this new environmental and contextual perspective of communication and interaction, all previous dimensions collapse to give rise to a proleptic world in which the new human emerges with transformational traits and habits. The *Homo Habilis* that we have been so far could soon be replaced by *Homo Prospectus*.

Led by future (not past)

We are only at the beginning of a development of which we can only glimpse, for now, a first horizon of possibilities. We were confronted precisely with the genesis of this new world – which includes software code, sensor networks, data and algorithms, artificial intelligence and platforms. But it is necessary to sound a last warning about the widespread use of the *post* prefix that characterizes many of the common trends of contemporary thought.

Post-industrial, post-human, post-modern, post-digital are the adjectives that are frequently used to connote the era we are living in. And I, too, to venture a bit of self-criticism, I have used the expression post-experiential.

I believe, however, that the time has come to use a new prefix in our discourses and reflections. In fact, I believe that we are not experiencing the end of an era, but the beginning of a new one. If this is true, then it is more appropriate to start using the *proto* prefix. We are not simply living the final or conclusive phase of an era. Rather, we are at the beginning of a new phase of the Anthropocene, a new transformative stage of cultural, social, technological, economic, organizational and business models and practices.

This shift of prefix from *post* to *proto* marks our ability to look at the "proto-data world" as the starting point of a new techno-human and socio-economic age and not simply as the closing phase of a bygone season. Mikhail Epstein reminds us of the true meaning of this prefix in his *The Transformative Humanities: A Manifesto*.

Proto means opening oneself to a potential future rather than merely following a sequential timeline, an emergence of possibilities more than a present stage of a predestined future. We are not to live *after* something (post-industrial age, post-modern, post-human condition, post-digitality).

We will no longer live as residues of the past, but we will be instantiated by the future. A future in which a philosophy of digital, of artificial and of synthetic will have the task of illuminating, with ethical foundations, the path of this new *Homo Prospectus*.

Afterword

by *Derrick de Kerckhove**

In short: sensing is not just measuring and understanding an existing given reality, it is creating and projecting a new reality. It is not just a matter of preventive diagnostics (medical, industrial and so on) but of a new, proactive techno-ecology, oriented towards the future thanks to the data.
Cosimo Accoto, this book, The Sensor

I teach a course on the anthropology of communication at the Politecnico of Milan. I feel fortunate that I was posted to the School of Design of that great institution because for one, my 50 odd students come from all over the world. So, instead of giving my class in Italian, I do it in English, which in Italy for me is rare and fun.

The other reason for my enjoyment is that these are design students, which means that, having passed advanced entrance exams, not only can they think, but they can also imagine. So I am more or less always on the look-out for material that can stimulate the imagination of my students. This is where *Il mondo dato* comes in.

I had not heard or read about Cosimo Accoto before his book appeared among others on my desk at *Media Duemila* and the title made me grab it. That title (in Italian "the world of/as data") in a flash had somehow made

* Derrick de Kerckhove is former Director of the McLuhan Program in Culture & Technology at the University of Toronto. Presently, scientific director of the Osservatorio TuttiMedia and of the Rome-based monthly *Media Duemila*, he is author of more than a dozen books translated into over ten languages. He is also Research Director at the Interdisciplinary Internet Institute (IN3) at the Universitat Oberta de Catalunya in Barcelona and teaches anthropology of communication at the Politecnico of Milan.

me already half-guess what the book was about. I did as we all do, that is, jump to the table of contents, and it was all there, all the issues, prospects and concerns that had been nagging at the back of my mind since the brusque acceleration of digital culture initiated by the mere concept of Big Data (buzzwords drive the digital economy). So the next thing was obviously to read the whole thing. I did that in two ways, first zigzagging between chapters, to get a feel for where it was leading, and then, reading the book very slowly several times to translate it.

The first read was accompanied, if not guided, by the idea that "this is a book my students need to read". That, of course, meant that I would have to translate it for them because more than half the class either does not speak, or read or even understand Italian. Translating is a job and I have already enough of those. But besides wanting to push my students, something else was driving me. It was the thought that this was a book I needed to understand in depth, and to do that there is nothing better than trying to translate it. I had learned that lesson from translating Marshall McLuhan's *From Cliché do Archetype* from English to French. It had been necessary to translate McLuhan in order to really understand him. Maybe the same would happen with translating Accoto.

What was it that I needed to understand? It is how electricity and its by-products were not only lighting and heating the environment of humans or transporting their voices and images but, by going digital, were penetrating the stuff of being, something I was already aware of, but quite incapable of formulating. Translating the book gave me renewed access as well as new coherence to the massive convergence of digital technologies readying itself to modify and regulate not only daily life and the economy but deep down, the essence of society. And, in fact, reading the text has been more than a cognitive experience, almost a physical one, as if the synthesis of data, platforms, AI and algorithms was in and out of our bodies, as if the whole environment was already, or soon destined to become suffused with this structuring entity, the full externalization of our central nervous system that McLuhan announced.

Hello new world indeed! What I am trying to work out in this monumental but sneaky social transformation is the ethics and the politics of a society soon to be completely straightjacketed by the clasp of technology. Blockchain or not, each one of us, the world over, will be traced, social-credited and rated down to the last detail. How does one behave under such conditions of vulnerability? What kind of political system can find legitimation in a transparent society? In order to answer that kind of question, you need to get the big picture.

What is so valuable about *In Data Time and Tide* is that it makes this vision possible. Instead of groping in the dark trying to tie together all the fragments of our digital culture that is smashing us to bits, we can see its ground and its constituent parts in the book. Henceforth we can appreciate that we occupy three kinds of integrated spaces: the physical one we are born into, the mental one that is our mind and now, the virtual, or more generally the digital one, which humanity occupies in a paradoxical way, both as acting agent and as source of an infinity of coded representations. The big idea here is that, thanks to sensors and actuators, the technological thrust is taking control not only of space but also of time, meanwhile changing the principal vector of civilization, that is, from past or present to future orientation and from correcting to formulating its course.

Accoto expands on this theme in the chapter The Datum, where he addresses the three fundamental coordinates of human experience, that is, time, space and self or (as it is designed in the book) subjectivity.

> New technologies engineer this proleptic and anticipatory orientation of our consciousness. In a more general sense, there is a real "precognitive vocation" of 21st century technologies, a push to the anticipated time that is effected by sensors and new data processing technologies instructed by artificial intelligence.
> Cosimo Accoto, this book, The Datum

The programming enabled in the third space also brings about a monumental change in the vector of time.

The book is really a philosophical explorative map of the "programmable" world created by code, data and algos. The notion of programming the world brings me back to translating McLuhan who, in *From Cliché to Archetype* observed that:

> Since Sputnik and the satellites, the planet is enclosed in a man made environment that ends "Nature" and turns the globe into a repertory theater to be programmed.
> McLuhan, *From Cliché to Archetype*, 1970

The programming links the hidden life of software and coding, the emergence of a new artificial data sensorium, the pervasive power of algorithmic intelligence, the shift from a present archival to a future oracular society, the world as a platform and the programmable "everything" from money, to law, to matter and beyond. Accoto surprisingly lights and links up unknown territories.

The book explores those territories in detail and in some fashion maps them out. Accoto is an explorer who knows how to describe his jungle.

I am grateful to him and to his publisher for having accepted my request to do this translation and to share it with my students who, by the way, were sufficiently inspired to make several critical presentations in class highlighting and commenting aspects that connected with their notions about design and culture. Many reported to me having been awoken to a new awareness about how urgent was this kind of investigation. One of them let me know that reading the book changed his specialization in order to concentrate on the study of communications.

References

Abend, P. and Fuchs, M. (2016), "The Quantified Self and Statistical Bodies", *Digital Culture & Society*, 2(1), pp. 5-21.

Acemoglu, D. and Restrepo, P. (2018), *Artificial Intelligence, Automation and Work*, MIT, Department of Economics, working paper.

Agrawal, A., Gans, J. and Goldfarb, A. (2016), «The Simple Economics of Machine Intelligence», *Harvard Business Review*, 17 November.

Agraval, A., Gans, J. and Goldfarb, A. (2018), *Prediction Machines. The Simple Economics of Artificial Intelligence*, Boston, Harvard Business Press.

Alur, R. (2016), *Principles of Cyber-Physical Systems*, Cambridge, MA, The MIT Press.

Amoore, L. and Piotukh, V. (eds.) (2016), *Algorithmic Life. Calculative Devices in the Age of Big Data*, New York, Routledge.

Andrejevic, M. and Burdon, M. (2015), «Defining the Sensor Society», *Television & New Media*, 16(1), pp. 19–36.

Antonopoulus, A.M. (2016), *Mastering Bitcoin: Unlocking Digital Crypto-Currencies*, Sebastopol, O'Reilly Media.

Arbesman, S. (2016), *Overcomplicated: Technology at the Limits of Comprehension*, New York, Penguin.

Arntz, M., Gregory, T. and Zierahn, U. (2016), «The Risk of Automation for Jobs in OECD Countries: A Comparative Analysis», *OECD Social, Employment and Migration*, Working Papers, 189, Paris, OECD.

Arstila, V. and Lloyd, D. (eds.) (2016), *Subjective Time: The Philosophy, Psychology and Neuroscience of Temporality*, Cambridge, The MIT Press.

Ash, J. (2018), *Phase Media: Space, Time and Politics of Smart Objects*, New York, Bloomsbury.

Ashford Lee, E. (2017), *Plato and the Nerd: The Creative Partnership of Humans and Technology*, Cambridge, The MIT Press.

Auerswald, P. (2017), *The Code Economy: A Forty-Thousand-Year Story*, Oxford, Oxford University Press.

Aukstakalnis, S. (2016), *Practical Augmented Reality: A Guide to the Technologies, Applications, and Human Factors for AR and VR*, Boston, Addison-Wesley.

Bernhardt, C. (2016), *Turing's Vision: The Birth of Computer Science*, Cambridge, The MIT Press.

Berry, D.M. (2011), *The Philosophy of Software. Code and Mediation in the Digital Age*, New York, Palgrave Macmillan.

Berson, J. (2016), *Computable Bodies. Instrumented Life and the Human Somatic Niche*, New York, Bloomsbury.

BI Intelligence (2016), *The Internet of Everything*, Research Report.

Boden, M.A. (2016), *AI: Its Nature and Future*, Oxford, Oxford University Press.

Boellstorff, T. and Maurer, B. (eds..) (2015), *Data, Now Bigger and Better!*, Chicago, Prickly Paradigm Press.

Bostrom, N. (2014), *Superintelligence: Paths, Dangers, Strategies*, Oxford, Oxford University Press.

Bratton, B.H. (2016), *The Stack: On Software and Sovereignty*, Cambridge, The MIT Press.

Brodmerkel, S. and Carah, N. (2016), *Brand Machines, Sensory Media and Calculative Culture*, London, Palgrave Macmillan.

Brunton, F. and Nissenbaum, H. (2015), *Obfuscation: A User's Guide for Privacy and Protest*, Cambridge, The MIT Press.

Brynjolfsson, E. and McAfee, A. (2014), *The Second Machine Age: Work, Progress, and Prosperity in a Time of Brilliant Technologies*, New York, W.W. Norton.

Bullynck, M. (2016), «Histories of Algorithms: Past, Present and Future», *Historia Mathematica*, 43(3), pp. 332–341.

Burniske, C. and Tatar, J. (2018), *CryptoAssets. The Innovative Investor's Guide to Bitcoin and Beyond*, New York, McGraw Hill.

Burrell, J. (2016), «How the Machine 'Thinks': Understanding Opacity in Machine Learning Algorithms», *Big Data & Society*, January, pp. 1–12.

Casey, M. and Vigna, P. (2018), *The Truth Machine: The Blockchain and the Future of Everything*, New York, St. Martin's Press.

Chamayou, G. (2016), *A Theory of Drone*, New York, The New Press.

Cheney-Lippold, J. (2018), *We Are Data: Algorithms and the Making of Digital Selves*, New York, New York University Press.

Choudary, S.P. (2015), *Platform Scale: How an Emerging Business Model Helps Startups Build Large Empires with Minimum Investment*, Singapore, Platform Thinking Labs.

Chun, W.H.K. (2011), *Programmed Visions: Software and Memory*, Cambridge, The MIT Press.

Chun, W.H.K. (2016), *Updating to Remain the Same: Habitual New Media*, Cambridge, The MIT Press.

Claire, L. and Reiller, B. (2016), *Open for Business. Harnessing the Power of Platform Ecosystems*, New York, Taylor & Francis.

Cox, G. (2013), *Speaking Code. Coding as Aesthetic and Political Expression*, Cambridge, The MIT Press.

Davenport, T.H. and Kirby, J. (2016), *Only Humans Need Apply: Winners & Losers in the Age of Smart Machines*, New York, Harper Collins.

Day, R.E. (2014), *Indexing It All: The Subject in the Age of Documentation, Information and Data*, Cambridge, The MIT Press.

Daugherty, P. and Wilson, J. (2018), *Human + Machine: Reimagining Work in the Age of AI*, Cambridge, Harvard Business Press.

De Filippi, P. and Wright, A., (2018), *Blockchain and the Law: The Rule of Code*, Boston, Harvard Business Press.

DeLanda, M. (2011), *Philosophy and Simulation: The Emergence of Synthetic Reason*, New York, Continuum.

Domingos, P. (2015), *The Master Algorithm: How the Quest for the Ultimate Learning Machine Will Remake Our World*, New York, Basic Books.

Dourish, P. (2016), «Algorithms and Their Others: Algorithmic Culture in Context», *Big Data & Society*, July-December, pp. 1–11.

Dourish, P. (2017), *The Stuff of Bits: An Essay on the Materialities of Information*, Cambridge, The MIT Press.

Durham, P.J. (2015), *The Marvelous Cloud: Toward A Philosophy of Elemental Media*, Chicago, Chicago University Press.

Epstein, M. (2012), *The Transformative Humanities: A Manifesto*, New York, Bloomsbury.

Evans, D. and Schmalensee, R. (2016), *Matchmakers. The New Economics of Multisided Platforms*, Cambridge, Harvard Business Review Press.

Evens, A. (2016), *Logic of the Digital*, London, Bloomsbury.

Ezrachi, A. and Stucke, M.E. (2016), *Virtual Competition: The Promise and Perils of the Algorithm-Driven Economy*, Cambridge, Harvard University Press.

Farman, J. (2012), *Mobile Interface Theory: Embodied Space and Locative Media*, New York, Routledge.

Farman, J. (ed.) (2015), *Foundations of Mobile Media Studies: Essential Texts on the Formation of a Field*, New York, Routledge.

Finn, E. (2017), *What Algorithms Want: Imagination in the Age of Computing*, Cambridge, The MIT Press.

Flach, P. (2012), *Machine Learning: The Art and Science of Algorithms that Make Sense of Data*, Cambridge, Cambridge University Press.

Flasinski, M. (2016), *Introduction to Artificial Intelligence*, London, Springer.

Floridi, L. (2012), «Big Data and Their Epistemological Challenge», *Philosophy and Technology*, 25(4), pp. 435–437.

Floridi, L. (2014), *The Fourth Revolution: How the Infosphere Is Reshaping Human Reality*, Oxford, Oxford University Press.

Frabetti, F. (2012), «Technology Made Legible. Software as a Form of Writing in Software Engineering», in Brillenburg Wurth (ed.), *Between Page and Screen: Remaking Literature Through Cinema and Cyberspace*, New York, Fordham University Press, pp. 157–170.

Frabetti, F. (2015), *Software Theory: A Cultural and Philosophical Study*, London, Rowman & Littlefield.

Franklin, S. (2015), *Control: Digitality as Cultural Logic*, Cambridge, The MIT Press.

Fuller, M. (ed.) (2008), *Software Studies: A Lexicon*, London, The MIT Press.

Fuller, M. and Goffey, A. (eds.) (2012), *Evil Media*, Cambridge, The MIT Press.

Fuller, M. and Goffey, A. (2012), «Algorithms», in their *Evil Media*, Cambridge, The MIT Press, pp. 69–82.

Gabrys, J. (2016), *Program Earth: Environmental Sensing Technology and the Making of a Computational Planet*, Minneapolis, University of Minnesota Press.

Galloway, A. and Thacker, E. (2007), *The Exploit: A Theory of Networks*, Minneapolis, University of Minnesota Press.

Galloway, A. (2006), *Protocol: How Control Exists After Decentralization*, London, The MIT Press.

Gehl, R. (2014), *Reverse Engineering Social Media: Software, Culture, and Political Economy in New Media Capitalism*, Philadelphia, Temple University Press.

Gehl, R. and Bakardijeva, M. (eds.) (2016), *Socialbots and Their Friends: Digital Media and the Automation of Sociality*, New York, Routledge.

Gillespie, T. (2016), «Algorithm», in B. Peter (ed.), *Digital Keywords: A Vocabulary of Information Society and Culture*, Princeton, Princeton University Press, pp. 18–30.

Gitelman, L. (ed.) (2013), *'Raw Data' is an Oxymoron*, Cambridge, The MIT Press.

Goodfellow, I., Bengio, Y. and Courville A. (2016), *Deep Learning*, Cambridge, The MIT Press.

Hajian, S., Bonchi, F., Castillo C. (2016), *Algorithmic Bias: From Discrimination Discovery to Fairness-Aware Data Mining*, KDD Tutorial.

Hansen, M.B.N. (2006), *New Philosophy for New Media*, London, The MIT Press.

Hansen, M.B.N. (2012), «Ubiquitous Sensation or the Autonomy of the Peripheral: Towards an Atmospheric, Impersonal and Microtemporal Media», in U. Ekman (ed.), *Throughout: Art and Culture Emerging with Ubiquitous Computing*, Cambridge, The MIT Press, pp. 63–88.

Hansen, M.B.N. (2015), *Feed-Forward: On the Future of Twenty-First Century Media*, Chicago, The Chicago University Press.

Hardjono, T., Shrier, D. and Pentland, A. (2016), *Trust. Data: A New Framework for Identity and Data Sharing*, Boston, Visionary Future.

Harney, S. (2014), «Istituzioni algoritmiche e capitalismo logistico», in M. Pasquinelli (ed.), *Gli Algoritmi del Capitale*, Verona, Ombre Corte, pp. 116–129.

Hendler, J. and Mulvehill, A.M. (2016), *Social Machines. The Coming Collision of Artificial Intelligence, Social Networking and Humanity*, New York, APress.

Hildebrandt, M. and Rouvroy, A. (eds.) (2013), *Law, Human Agency and Autonomic Computing: The Philosophy of Law Meets the Philosophy of Technology*, Abingdon, Routledge.

Hildebrandt, M. and de Vries, K. (eds.) (2013), *Privacy, Due Process and the Computational Turn: The Philosophy of Law Meets the Philosophy of Technology*, Abingdon, Routledge.

Hildebrandt, M. (2015), *Smart Technologies and the End(s) of Law: Novel Entanglements of Law and Technology*, Cheltenham, Edward Elgar.

Hildebrandt, M. and van den Berg, B. (eds.) (2016), *Information, Freedom and*

Property. The Philosophy of Law Meets the Philosophy of Technology, Abingdon, Routledge.

Hoelzl, I. and Rémi, M. (2015), *Softimage: Towards a New Theory of the Digital Image*, Chicago, The University of Chicago Press.

Hommels, A. and Mesman, J., Bijker, W.E. (eds.) (2014), *Vulnerability in Technological Cultures: New Directions in Research and Governance*, Cambridge, The MIT Press.

Hong, S.H. (2016), «Data's Intimacy: Machinic Sensibility and the Quantified Self», *Communication+1, 5, Machine Communication*, pp. 1–36.

Hu, T.H. (2015), *A Prehistory of the Cloud*, Cambridge, The MIT Press.

Hui, Y. (2016), *On the Existence of Digital Objects*, Minneapolis, University of Minnesota Press.

Janssens, L. (ed.) (2016), *The Art of Ethics in the Information Society*, Amsterdam, Amsterdam University Press.

Jarzombek, M. (2016), *Digital Stockholm Syndrome in the Post-Ontological Age*, Minneapolis, University of Minnesota Press.

Jones, C.A., Mather, D. and Uchill, R. (2016), *Experience: Culture, Cognition and Common Sense*, Cambridge, The MIT Press.

Johnston, J. (2010), *The Allure of Machinic Life: Cybernetics, Artificial Life and the New AI*, Cambridge, The MIT Press.

Katz, M.B. (2015), *Make it New: The History of Silicon Valley Design*, Cambridge, The MIT Press.

Kelly, K. (2016), *The Inevitable: Understanding the 12 Technological Forces That Shape Our Future*, New York, Viking Penguin.

Kitchin, R. and Dodge, M. (2011), *Code/Space: Software and Everyday Life*. Cambridge, The MIT Press.

Kitchin, R. and Perng, S.Y. (eds.) (2016), *Code and the City*, New York, Routledge.

Kitchin, R. (2014), *The Data Revolution*, London, Sage Publications.

Kurzweil, R. (2005), *The Singularity is Near. When Humans Transcend Biology*, London, Penguin.

Laney, D. (2018), *Infonomics: How to Monetize, Manage, and Measure Information as An Asset for Competitive Advantage*, New York, Bibliomotion.

Lecavalier, J. (2016), *The Rules of Logistics: Wallmart and the Architecture of Fulfillment*, Minneapolis, University of Minnesota Press.

Letouzé, E. and Sangokoya, D. (2015), *Leveraging Algorithms for Positive Disruption: On Data, Democracy, Society and Statistics*, Data-Pop Alliance, working paper.

Libert, B., Beck, M. and Wind, J. (2016), *The Network Imperative: How to Survive and Growth in the Age of Digital Business*, Boston, Harvard Business Review Press.

Lipson, H. and Kurman, M. (2016), *Driverless: Intelligent Cars and the Road Ahead*, Cambridge, The MIT Press.

Lupton, D. (2016), *The Quantified Self*, Cambridge, Polity Press.

Mackenzie, A. (2006), *Cutting Code: Software and Sociality*, Oxford, Peter Lang.

Mackenzie, A. (2010), *Wirelessness: Radical Empiricism in Network Cultures*, Cambridge, The MIT Press.

Mackenzie, A. (2015), «The Production of Prediction: What Does Machine Learning Want?», *European Journal of Cultural Studies*, 18(4/5), pp. 429–445.

Mackenzie, A. (2013), «Programming Subjects in the Regime of Anticipation: Software Studies and Subjectivity», *Subjectivity* 6(4), pp. 391-405.

Mayer-Schönberger, V. and Ramge, T. (2018), *Reinventing Capitalism in the Age of Big Data*, New York, Basic Books.

Manovich, L. (2001), *The Language of New Media*, Cambridge, The MIT Press.

Manovich, L. (2013), *Software Takes Command*, New York, Bloomsbury.

Marr, B. (2016), *Big Data in Practice*, Chichester, Wiley.

Markoff, J. (2015), *Machines of Loving Grace: The Quest for Common Ground Between Humans and Robots*, New York, Harper Collins.

McCullough, M. (2013), *Ambient Commons: Attention in the Age of Embodied Information*, Cambridge, The MIT Press.

McKinsey (2016), «The CEO Guide to Customer Experience», *McKinsey Quarterly*, August.

McTear, M., Callejas, Z. and Griol, D. (2016), *The Conversational Interface: Talking to Smart Devices*, London, Springer.

Miyazaki, S. (2012), «Algorhythmics: Understanding Micro-Temporality in Computational Cultures», *Computational Cultures Journal*.

Moazed, A. and Johnson, N.L. (2016), *Modern Monopolies: What It Takes to Dominate the 21st Century Economy*, New York, St. Martin's Press.

Mosco, V. (2014), *To the Cloud. Big Data in a Turbulent World*, Boulder, Paradigm Publishers.

Mougayar, W. (2016), *The Business Blockchain: Promise, Practice and Application of the Next Internet Technology*, Hoboken, Wiley.

Munoz C., Smith M. and Patil D.J. (2016), *Big Data: A Report on Algorithmic Systems, Opportunity and Civil Rights*, Executive Office of the President of USA, Washington.

Nadin, M. (ed.) (2016), *Anticipation and Medicine*, Cham, Springer.

Nafus, D. (ed.) (2016), *Quantified: Biosensing Technologies in Everyday Life*, Cambridge, The MIT Press.

National Science and Technology Committee (2016), *Preparing for the Future of Artificial Intelligence*, Executive Office of the President of USA, Washington.

Neff, G. and Nafus, D. (2016), *The Quantified Self*, Cambridge, The MIT Press.

Parisi, L. (2013), *Contagious Architecture: Computation, Aesthetics and Space*, Cambridge, The MIT Press.

Pariser, E. (2012), *The Filter Bubble: How the New Personalized Web Is Changing What We Read and How We Think*, New York, Penguin Books.

Parker, G.G., Van Alstyne, M.W. and Choudary, S.P. (2016), *Platform Revolution: How Networked Markets Are Transforming the Economy and How to Make Them Work for You*, New York, Norton & Company.

Pasquale, F. (2015), *The Black Box Society: The Secret Algorithms that Control Money and Information*, Cambridge, Harvard University Press.

Pasquinelli, M. (ed.) (2014), *Gli algoritmi del capitale. Accelerazionismo, macchine della conoscenza e autonomia del comune*, Verona, Ombrecorte.

Pentland, A. (2014), *Social Physics: How Good Ideas Spread: The Lessons from a New Science*, New York, Penguin Press.

Pentland, A., Shrier, D., Hardjono, T. and Wladawsky-Berger, I. (2016), *Towards an Internet of Trusted Data: A New Framework for Identity and Data Sharing*, Boston, MIT Connection Science, working paper.

Picard, R.W. (2007), *Affective Computing*, Cambridge, The MIT Press.

Pieraccini, A. (2012), *The Voice in the Machine*, Cambridge, The MIT Press.

Pine, B.J. and Gilmore, J.H. (1999), *The Experience Economy: Work is Theatre & Every Business a Stage*, Boston, Harvard Business School Press.

Pine, B.J. and Korn, K.C. (2011), *Infinite Possibility: Creating Customer Value on the Digital Frontier*, San Francisco, Berrett-Koehler Publishers.

Pink, S., Ardèvol, E. and Lanzeni, D. (eds.) (2016), *Digital Materialities: Design and Anthropology*, London, Bloomsbury.

Platoni, K. (2015), *We Have the Technology: How Biohackers, Foodies, Physicians and Scientists Are Transforming Human Perception, One Sense at a Time*, New York, Basic Books.

Preston, C. (2018), *The Synthethic Age: Outdesigning Evolution, Resurrecting Species, and Reengineering Our World*, Cambridge, The MIT Press.

Pschera, A. (2016), *Animal Internet: Nature and the Digital Revolution*, New York, New Vessel Press.

Portanova, S. (2013), *Moving Without a Body: Digital Philosophy and Choreographic Thought*, Cambridge, The MIT Press.

Postolache, O.A., Mukhopadhyay, S.C., Jayasundera, K.P. and Swain, A.K. (2016), *Sensors for Everyday Life: Healthcare Settings*, Cham, Springer.

Raval, S. (2016), *Decentralized Applications: Harnessing Bitcoin's Blockchain Technology*, Sebastopol, O'Reilly.

Rajlich, V. (2012), *Software Engineering: The Current Practice*, Boca Raton, CRC Press.

Ratti, C. and Claudel, M. (2016), *The City of Tomorrow: Sensors, Networks, Hackers, and the Future of Urban Life*, New Haven, Yale University Press.

Raunig, G. (2016), *Dividuum: Machinic Capitalism and Molecular Revolution*, Cambridge, The MIT Press.

Rid, T. (2016), *The Rise of the Machines: A Cybernetic History*, New York, Norton & Company.

Rosenbloom, P.S. (2012), *On Computing: The Fourth Great Scientific Domain*, Cambridge, The MIT Press.

Ross, A. (2016), *The Industries of the Future*, New York, Simon & Schuster.

Rossiter, N. (2016), *Software, Infrastructure, Labor*, New York, Routledge.

Roth, A.E. (2016), *Who Gets What – and Why: The New Economics of Matchmaking and Market Design*, Boston, Eamon Dolan Book.

Sadin, E. (2011), *La société de l'anticipation*, Paris, Incult Essais.

Schwab, K. (2016), *The Fourth Industrial Revolution*, Geneva, World Economic Forum.

Schwartz, S.I. (2015), *Street Smart. The Rise of Cities and the Fall of Cars*, New York, Perseus Books.

Scott Barker, T. (2012), *Time and the Digital: Connecting Technology, Aesthetics and a Process Philosophy of Time*, Hanover, Dartmouth College Press.

Scott, L. (2016), *The Four-Dimensional Human: Ways of Being in the Digital World*, London, William Heinemann.

Sejnowski, T. (2018), *The Deep Learning Revolution*, Cambridge, MIT Press.

Selke, S. (ed.) (2016), *Lifelogging: Digital Self-Tracking and Lifelogging, between Disruptive Technology and Cultural Transformation*, London, Springer.

Seligman, M.E.P., Railton P., Baumeister, R.F. and Sripada, C. (2016), *Homo Prospectus*, Oxford, Oxford University Press.

Seyfert, R. and Roberge, J. (eds.) (2016), *Algorithmic Cultures: Essays on Meaning, Performance and New Technologies*, Abingdon, Routledge.

Shanahan, M. (2015), *The Technological Singularity*, Cambridge, The MIT Press.

Stiegler, B. (2017), *Automatic Society: The Future of Work*, Cambridge, Polity Press.

Strengers, Y. (2016), «Envisioning the Smart Home: Reimagining a Smart Energy Future», in S. Pink, E. Ardèvol and D. Lanzeni (eds.), *Digital Materialities. Design and Anthropology*, London and Oxford, Bloomsbury Academic.

Shaughnessy, H. (2015), *Shift: A Leader's Guide to Platform Economy*, Boise, Tru Publishing.

Shaughnessy, H. (2016), *Platform Disruption Wave: How the Platform Economy is Changing the World*, Boise, Tru Publishing.

Shaviro, S. (2014), *The Universe of Things: On Speculative Realism*, Minneapolis, Minnesota University Press.

Shaviro, S. (2016), *Discognition*, London, Repeater Books.

Sheller, M. (2016), «Mobilizing the New Mobilities Paradigm», in *Applied Mobilities*, pp. 10–25.

Shrier, D. and Pentland, A. (eds.) (2016), *Frontiers of Financial Technologies: Expeditions in Future Commerce, from Blockchain and Digital Banking to Prediction Markets and Beyond*, Boston, Visionary Future.

Shrobe, H., Shrier, D. and Pentland, A. (ed.) (2018), *New Solutions for Cybersecurity*, Cambridge, The MIT Press.

Siegel, E. (2013), *Predictive Analytics*, Hoboken, Wiley.

Silver, N. (2012), *The Signal and the Noise: The Art and Science of Prediction*, London, Penguin.

Srnicek, N. (2016), *Platform Capitalism*, Hoboken, Wiley.

Stevens, T. (2016), *Cyber Security and the Politics of Time*, Cambridge, Cambridge University Press.

Sundararajan, A. (2016), *The Sharing Economy: The End of Employment and the Rise of Crowd-Based Capitalism*, Cambridge, The MIT Press.

Sundbo, J. and Sorensen, F. (eds.) (2013), *Handbook on the Experience Economy*, Cheltenham, Edward Elgar Publishing.

Sussna, J. (2015), *Designing Delivery: Rethinking IT in the Digital Service Economy*, Boston, O'Reilly Media.

Swan, M. (2016), «Blockchain Temporality: Smart Contract Time Specifiability with Blocktime», in J.J. Alferes, L. Bertossi, G. Governatori, P. Fodor and D. Roman (eds.), *Rule Technologies: Research, Tools, and Applications*, 9718, pp. 184–196.

Swanstrom, E. (2016), *Animal, Vegetable, Digital: Experiments in the New Media Aesthetics and Environmental Poetics*, Tuscaloosa, The University of Alabama Press.

Tapscott, D. and Tapscott, A. (2016), *Blockchain Revolution. How the Technology Behind Bitcoin is Changing Money, Business and The World*, New York, Portfolio/Penguin.

Tucker, P. (2014), *The Naked Future: What Happens in a World that Anticipates Your Every Move?*, New York, Current Penguin.

Yuill, S. (2008), «Interrupt», in M. Fuller (ed.), *Software Studies: A Lexicon*, Cambridge, The MIT Press.

Wajcman, J. (2015), *Pressed for Time: The Acceleration of Life in Digital Capitalism*, Chicago, University of Chicago Press.

Warwick, K. and Shah, H. (2016), *Turing's Imitation Game: Conversations with the Unknown*, Cambridge, Cambridge University Press.

World Economic Forum (2016), *A Blueprint for Digital Identity: The Role of Financial Institutions in Building Digital Identity*, Working Paper.